The Junkman's Guide

The Junkman's Guide

(Published in hardcover as *Farmer's Law*)

RICHARD N. FARMER

A SCARBOROUGH BOOK
STEIN AND DAY/*Publishers*/New York

FIRST SCARBOROUGH BOOKS EDITION 1978
The Junkman's Guide was originally published in hardcover
by STEIN AND DAY/*Publishers* as *Farmer's Law*.

Copyright © 1973 by Richard N. Farmer
Library of Congress Catalog Card No. 73-80843
All rights reserved
Designed by David Miller
Printed in the United States of America
Stein and Day/*Publishers*/Scarborough House,
Briarcliff Manor, N.Y. 10510
ISBN 0-8128-2469-5

To George A. Farmer,
who taught me all about junk,
and lots of other things

Preface

The junk readers I have known seem to be a step or two ahead of everyone else most of the time. Things that appear very obscure to most people are obvious to them—a good junk reader knows what is going on in a city, neighborhood, or country long before others have figured out what the problem is. When a good junk reader moves into a new neighborhood, he would know within a week, before he met anyone, that Doc Jones was pretty careless about tools; that Sam was a frugal, cautious man who rarely did anything rash; that Mrs. Gillman drank a bit too much; and that one of the kids in the house two doors down was probably a real genius.

When junk readers go into business or try to sell something, they invariably succeed. They know what kinds of skilled men are around, and what they cost, no matter what strange country they are in. A good junk reader told me more about what was really happening in Japan and southern Italy than anyone, long before any statistics were published; another one gave me a very accurate account of Sicily after being there for only a few hours. Indeed, junk readers, like witches, seem to be able to do all sorts of mysterious things that no one else can do.

I discovered long ago that most junk readers are naturals, since no one ever ran a course or wrote a book about such matters. Somehow they just *know*. It was only when a few students and I began talking about the art that we realized that it could be taught. Junk reading is not witchcraft, but rather a science, which is what this book is about.

Anyone who spends time staring at old piles of rusty iron or poking through piles of used bottles is regarded as eccentric, so few junk readers publicize their exploits much. They just go on winning.

Moreover, I discovered, once I began to examine this whole problem, that no one in power, and no one involved in large-scale planning, had ever read junk either, which was one reason why nothing ever worked out the way it was supposed to. If Mr. McGovern knew how to read junk, and had applied what he read, he might have received quite a few million more votes; if economic planners in Africa could read junk, many more of their carefully thought out plans would work; if salesmen could read junk, they could exceed their quotas any time; if hippies could read junk, they could drop out much more efficiently; if managers and businessmen could read junk, they might significantly cut personnel costs; if schoolmasters and teachers could read junk, they would reduce drop-out rates and do a better job of educating the young. This ability is a powerful tool for all sorts of people in all sorts of enterprises.

Hence this book. Junk secrets are revealed, for fun and profit. Those who know junk secrets are not the least bit eager to tell the world what they have learned—why spoil a good thing? But now we are in crisis. In the old days, when junk was not so prolific as it is now, it somehow disappeared without anyone getting fussed. Now we have serious citizens very concerned about ecology, pollution, and trash, and all sorts of good and bad suggestions are made daily about what to do about it. Since the people making all these suggestions don't know how to read junk, or even to analyze it economically, most ideas don't work too well. So, in the public interest (and for profit), I decided to tell all here. What follows, trivial as it may seem, is actually a clue to how the modern world really operates. Indeed, it *is* the way the modern world works, and he who understands it can live like a king, in any part of the world he chooses.

Thanks and acknowledgments are a part of a preface, and here three are necessary. Jeff Arpan, one of my brilliant doctoral students, originally urged me to write all this stuff up. Thanks, Jeff. Dick Lurie, editor of *P & I Planning,* thought enough of early efforts to publish an article on junk. And most of all, my father, George A. Farmer, deserves the most credit, because he is one of the very few creative junk users I have ever met, and certainly the best. What he can't do with a chunk of steel plate, a cast-off typewriter roll, an ancient alarm clock with the face missing, or anything else, isn't worth doing. If there were half a million others like him in the United States, this book couldn't have been written, because there wouldn't be any junk worth talking about. It would all be in valuable use.

Reading junk, like reading any other language, is subject to error. Hence any errors in judgment, misreadings of piles of trash around the world, and factual misstatements, all belong to me.

Richard N. Farmer

Contents

2	Introduction	15
2	Total Affluence: The United States	31
3	Junking the System	57
4	Making Money on Junk: Autos	83
5	Nothing Gets Used	100
6	Everything Gets Used	111
7	Piling It Up	128
8	Communist Junk: The Special Case	140
9	The Age of Affluence	151
10	What to Do About It	169

Farmer's Law on Junk

Farmer's Law: What Goes In, Comes Out

Corollary 1: He who sees what comes out and why, gains wisdom.

Corollary 2: He who sees only half the problem will be buried in the other half.

Corollary 3: One man's junk is another's income—and sometimes his priceless antique.

Corollary 4: 10,000 years from now, the only story this civilization will tell will be in its junk piles—so observe what is important!

Corollary 5: Seers and soothsayers read crystal balls to find the future. Less lucky men read junk—with more success.

Corollary 6: A rose is a rose is a rose, but junk is not junk is not junk. It never is quite what you think it is.

Corollary 7: Happiness at age ten was finding an empty six pack of returnable Coke bottles. The poor kids these days will never know what they missed, which is why we have a generation gap.

1

Introduction

This is a book about junk—rags, bottles, sacks, mountains of rusty scrap iron, seas of wrecked autos, rivers of used motor oil—the rich droppings of a modern civilization. We never seem to get rid of it fast enough. A garbage strike can paralyze a great city, and everywhere we look there are piles of trash. Few people do more with junk than try to get rid of it, or get mad at those officials who somehow can never dispose of it fast enough. Yet the study of junk is far from unpleasant or dull. Like the physician who asks us to fill that familiar urine sample bottle, the junk doctor studies the outflows of society to reveal its basic health or sickness, its prosperity or despair.

Junk studies focus on the other end of our civilization, the unpleasant end. Anything that goes in comes out, sooner or later, usually in some unwanted form. Often it is easier and cheaper to take a close look at the *out* part of the civilization than the *in* part. Such study can tell you more about what is really going on than anything else, in large part because junk doesn't lie. Propagandists and planners may fudge a bit in describing their culture, but no one knows or cares about junk, so it just sits there waiting to be examined for fun and profit.

The fun part of the problem is in understanding what a culture is like, and why. The profit part is using junk-reading techniques to make a living; figuring out what kinds of labor are available in a given place or country; discovering new life styles, which only are possible with judicious use of junk; and deciding what you need to function well in a country, no matter what you have in mind. As innumerable junkmen have discovered for a long time, there is real money in the leavings of society, but as far as I can tell, no one has used junk analysis for better living and greater success. And that is one major purpose of this book.

If you like to be one up on everyone, this book will tell you how. Serious scholars and marketing men will spend months, or even years trying to figure out what makes a given culture tick. If you read this book carefully and learn your lessons well, you can stroll casually around the neighborhood, peer at a few junk piles, examine (at a distance) some garbage, and within a few hours give a carefully reasoned analysis of what is really going on. If you're a salesman or manager, this quick information can easily be turned to a profit. If you want to drop out of society and find a new life style, such knowledge can be used to live well and easily. And if you're an economic development expert, junk study can show you exactly what is needed next year to keep the country growing fast.

Like a urine specimen, junk doesn't lie. Whatever you see, if you look in the right places at the right things, will be the truth, which is more than you can say about cocktail party conversation, company news releases, government information-service bulletins, or even statistical abstracts. In short, this book gives you a useful way to figure out the world you are in.

So we will meander around the globe through the world of junk. We may be the first laymen to do so—how often have you seen any junk pictured in any major magazine? Junk or garbage is simply not discussed by nice people. Oh, well, that's their problem—we can beat the game by studying junk piles of the world.

Since junk and other types of pollution are so unpopular these days, we also will take a look at what to do about it.

There is a logic to junk, which few have ever thought through: given a set of cultural, political, and economic forces, you can easily forecast exactly the kinds of junk a culture will have. And if you know what is going to happen, you can do something about it by changing some key elements of the problem. Most anti-junk and anti-pollution campaigns have trouble because few, if any, planners have bothered to work out the various laws of junk. Here, in addition to everything else, we have a model, beautifully structured, full of high-powered insights into what is really going on. From the model, we can easily solve anyone's junk problem, if they are willing to listen. Not bad for a short book, but true nonetheless.

Junk and the Businessman

For entrepreneurs and firms going as strangers to foreign countries, junk studies can be valuable. Junk tends to be a leading economic indicator, showing those who care to look what is going to be happening five or ten years from now. For overseas contractors and investors taking a close look at a society for the first time, careful study of trash in the country can be one of the more productive things a survey team can do. Among other things, junk surveys can suggest the skills and abilities of the work force, along with their relative honesty. Most economic data is erroneous, for reasons beyond the control of statisticians and ministries; junk studies can often correct defective data, as well as flesh out in considerable detail what is *really* going on in the countryside.

Thus one can read in the statistical abstract that a country has a per-capita income of $100 per year. What does this mean? Besides saying that the country is rather poor, it tells an investigator nothing about what is actually going on in the society. But if he wanders around scrap yards and notices that old tires are being carefully collected and being made into sandals, he can infer something about the skills of the work force. And if the sandals are attractively designed, he can deduce something about the creative instincts of local craftsmen.

Whether the tires are locked up and guarded twenty-four hours per day or just piled casually on the sidewalk says a great deal about standards of honesty among the workmen and in the country. A further useful point that can be inferred is what kinds of production are really going on in the country. Sandal makers with one or two employees rarely get counted in anybody's industrial census, yet such production may be a significant part of the total, to say nothing of providing sources of skilled labor unnoticed in anyone's report of formally educated people.

This author, in case anyone cares, has been a peripatetic junk watcher for twenty years, in ten different countries. Back in the 1930s, we saved string, boiled fats, and found old car parts down in creeks, because we had to. Fishing out a Model T Ford axle from a dry creek to use as the main support of a homemade drill press is probably a lost art in the United States, but it still goes on abroad.

In reading trash, one rarely is wrong. I built a whole labor policy around junk observations in one poor Middle Eastern country, and it worked better than anything else around.

Trash Cycles in Economic Development

Cultures and countries go through quite specific cycles of trash disposal and utilization, and some knowledge of how these cycles work can be helpful. To read the cycle, you need to get out to some provincial city, away from the political or commercial capital. This is necessary because there may be considerable distortion in the trash system around foreign embassies, which are usually located in the more affluent sections of cities. Foreigners tend to import their own cultural standards wherever they can, and local citizens try to emulate them. Thus in the capital, it is common to find such trash as bottles thrown away and forgotten. If the city has a high-income district somewhat removed from the rest of the city, the bottles will sit around for some time before anyone picks them up. Ras Beirut, in Lebanon, is one example of this situation, and downtown Tokyo is another. Casual observers may assume from such urban sectors that the national situation is just like home, which can be a serious error.

Moreover, if trash is handled in the Western manner anywhere, it will be at the center of the country's affairs, which is not what we are after. On the other hand, getting too far into the boondocks may also be misleading, even in the United States. What is needed is a nice average situation; and a medium-sized provincial city fills the bill.

Finding the junkyards is really easy. In most countries, there are only one to two decent hotels for foreigners in any medium-sized city. Start from there and wander into the commercial district, which invariably is close at hand. Keep an eye out for little alleys and narrow streets cutting off from this main district. Often these lead to what you're after. If these are not around, keep going to the edge of the district. It will start getting shabbier, and often more native as well. Instead of neon signs, we find hand-painted local-language placards. Stores displaying shiny typewriters from Olivetti and Japanese motor bikes give way to shops selling local foods and clothing. A few more blocks will get you to the secondhand stores, pawnshops, and real local dives, if they exist. Depending on the size of the city and the local culture, this is anywhere from five to fifteen blocks from your hotel.

Somewhat beyond the dives and such, the junkyards, if any, begin. They will always be tucked away someplace in a back alley, in a section where the land values are low. Some will have high fences; others will be open. And, as we shall see, in some places you won't find any, but you will find small workshops busily manufacturing stuff from junk. In these cases, the above instructions apply, but instead of a junkyard, you will stumble on someone's inventory of tires, old cans, oil drums, or what not.

Reading Junk

Once in the right location, your junk-reading career can begin.

There are two major kinds of junk: autos and everything else. Each will be considered in turn. The pattern runs about like Figure 1. But the real sophistication of the technique is in spotting the aberrations from the norm. Figure 1 shows only the general pattern. The person who can spot the unusual case, particularly if he is looking for a good spot to locate a manufacturing facility or a source of supply of good labor for his new contract, is the one who can keep his costs down and his efficiency way up.

Figure 2 shows the same pattern graphically, although again, the deviations are the interesting things to look for. This pattern, based on per-capita income in the country, runs about as follows:

1. Nothing gets used. This is the case of a really primitive society, where no one knows how to use or modify anything. It is a relatively rare case in modern times, but examples exist in such cultures as in the South Sea Islands, where natives simply have no skills at transforming anything. The Americans or Japanese might have left on these islands almost usable trucks and jeeps which are still there unless someone from outside has come to get them. As late as 1962, virtually all the locomotives and cars of the Hejaz Railway in Saudi Arabia were still sitting where Lawrence of Arabia stopped them in 1916. The nomads had taken all the wood, but they could not figure out how to use a forty-ton steam locomotive or a fifteen-ton freight car.

The dynamics of such situations should also be noted, since

Figure 1: Country Trash Profiles

Countries or Areas	Per Capita GNP Per Yr.	Observable Trash
South Sea Islands, Primitive, and remote African tribes, Desert nomads	under $50	Everything-nothing gets used.
Afghanistan, Senegal, Sudan, Tanganyika, India, Pakistan, Cambodia, Tunisia, Bolivia, Jordan, Nepal, Upper Volta, Yemen, etc.	$51–$200	Nothing or very little—everything gets used for some purpose, often in ingenious ways—autos totally stripped.
Argentina, Brazil, Mexico, Lebanon, Morocco, Jamaica, Greece, Malaya, etc.	$201–$500	Low-grade material—broken bottles, cement fragments, occasional tires (rare), leaking plastic bottles, and similar autos almost, but not quite totally stripped.
Japan, Venezuela, Israel, Finland, Iceland	$501–$1200	Beginnings of affluence: usable trash accumulates. Bottles, cans, tires, occasional abandoned cars, etc.
West Germany, United Kingdom, Switzerland, Norway, Australia, France, etc.	$1201–$2500	Trash problems becomes a public issue: what to do with it? Abandoned cars, throwaway containers, paper as a key problem.

countries tend to shift through time on the scale shown in Figure 1. Thus, by 1964, the Hejaz Railway was being rebuilt, and the junk was being picked up and used. As the culture develops, trash-use patterns change considerably, so one has to watch for new developments all the time

Per-capita incomes in these cultures are typically below $50 per year, and anyone working in such a situation might as well resign himself to importing everything, since the local culture will lack even rudimentary skills.

2. Everything gets used. As income rises a bit, and as local citizens' acquire skills, we find a culture where there are no junkyards. This is because everything, no matter how unlikely, gets reused almost immediately. The society is too poor to leave junk lying around, and the familiar Western pattern of old auto tires in creeks and beer bottles in the parks is nonexistent. People know how to use such stuff. An old beer bottle can serve as a drinking cup to a family with nothing else; and old tires are laboriously cut apart by hand to salvage the wire beads in the casing. In more advanced examples, the rubber becomes sandals, and old newspapers wrap packages.

Many parts of India and Pakistan are in this situation, along with other Asian countries. One additional reason everything gets used is that there is so little of everything. The average Indian family is not a user of disposable (or even returnable) soft-drink six-packs, nor do they own cars or other durable consumer goods. Newspaper consumption per capita is very low, which means that there is a scarcity, even for wrapping. Moreover, in such a country, there tends to be large-scale unemployment (or semiemployment), which means that it may well pay someone to take the time and effort to make even a very small income by running around and picking up virtually anything. A man making nothing may find that an income of 20 cents a day is high—so even finding four or five usable bottles per day is a pretty good deal.

Per-capita incomes in such cases are around $60 to $200

per year. As the society gets more skilled, we encounter several subphases here, as follows:

First, there is the "use-as-is" stage, as suggested above. But this is quickly followed by a system of crude reprocessing of all sorts of junk. Oil barrels get flattened out and used as housing materials (Saudi Arabia, West Indies), cans have handles soldered on to be used as pots, and so on. In some cases, such reprocessing can get quite artistic, as in the oil-drum bands in the West Indies, or in elaborate decorated sandals made from old tires.

Such crude reprocessing tends to evolve rather quickly into more sophisticated types of work. Thus the old cans become raw material for toys (Japan, circa 1950); autos are neatly dissected and steel plates are torched out of roofs and doors (Saudi Arabia, circa 1965); and so on. At this stage, auto junkyards begin to appear for the first time. Lebanon in the early 1960s was one example. In earlier cycles, there are not enough cars and trucks around to create junkyards, but now there are—and

Figure 2: Trash Problems

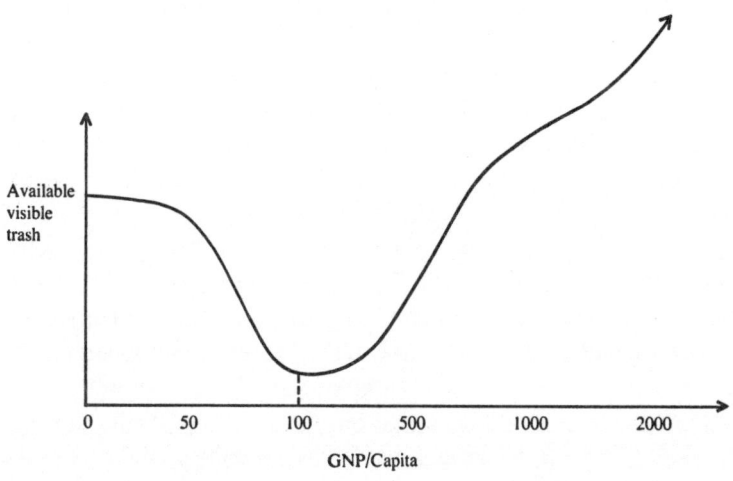

the junkyards are empty. That is, every conceivable usable nut and bolt is used, and what is left are the absolutely stripped hulks of cars and trucks. Note that in these countries, some skilled labor exists, so anything usable is removed and used. Moreover, labor is relatively cheap, so each nut, bolt, piece of glass, length of copper tubing, and so on in any vehicle is laboriously removed by hand and reused. A visitor to a Lebanese auto junkyard can be sure, when he sees those absolutely stripped hulks, that there is an ingenious and useful supply of mechanics and workers around.

Cars Old and New

Another visible cross-check on this point can be made by observing the kinds of cars being used in routine daily service. Many countries in this bracket have had serious exchange-control problems for decades, and they think nothing of running twenty-, thirty-, or even forty-year-old vehicles around in routine daily service. When one sees 1930 Chrysler tourings used as family cars (Uruguay, 1969); 1930 Nashes used as taxis (Egypt, 1965); or Model A Fords used as jitneys, making 80,000 miles or so a year (Peru, 1972); then one can conclude that not only will the auto junkyards be empty but also that there are some very ingenious mechanics, machinists, parts manufacturers, and tinkerers hidden away on back streets. (Try to buy a main bearing insert for your 1930 Chrysler in the United States to see what I mean; it isn't easy!).

Curiously, these are countries where we often hear complaints that there is no skilled labor; that firms can't find good men to do complex disciplined industrial tasks; or that somehow we can never seem to find the right kinds of craftsmen to do modern work. Could it be that we have been looking in the wrong places? Any culture that has the necessary skills to keep a 1930 Chrysler running must have something going for it, including a lot of very good technicians, organizers, and mechanics. A stroll down some back streets might pay big dividends to a labor recruiter looking for the right kinds of talent.

Few countries in this category have their own steel mills and, if they do, there is not enough scrap around to make its use worthwhile, so the very stripped hulks may lie around for a long time. But eventually someone makes a contract, and the hulks, which are nice pure steel, get sent to Japan or Europe for melting down.

Special situations occur here too, particularly if the Western firms have been active. Thus Saudi Arabia since 1960 has been a good source of steel exports, from the Eastern Province, since the various oil activities have generated a large amount of high-quality steel scrap. Careful examination of such scrap can also say a lot about society. In the early 1960s, scrap often contained lots of goodies, such as almost complete diesel-electric generating sets; truck engine blocks; barrels of nuts and bolts, unused; and the crane jibs. But as the local society develops, the quality of scrap goes down, and now the stuff being exported is just lumps and pieces of steel with little value beyond its scrap content. The Saudis are rapid learners, and they have found out how to use many useful odds and ends left over from the oil fields. If anyone had bothered to plot scrap composition in Arabia from 1955 to 1970, they would have found out more about the rapid development of labor skills in the economy than from any other economic indicator.

3. The piling-up phase. As the culture gets more affluent, it gets more stuff; and the stuff tends to pile up here and there. Labor, particularly skilled labor, is being drawn off into more productive pursuits, so cars begin to have bits and pieces of hard-to-get things left on them. Really useless stuff also begins to pile up, since even unskilled labor is now a bit expensive, and visitors can find chunks of concrete, broken glass, and other items scattered around here and there.

As this stage progresses, some useful stuff also begins to stack up in odd places. Newspapers can't all be used (note the imbalance—there is more packing materials around to use, along with lots more newspapers, as citizens read more). An occasional abandoned car is seen in an odd corner, and while most easily accessible stuff is gone (for example, tires, mechanical parts), some of the nuts and bolts are still on it. Odd bits and pieces of steel show up in junkyards, as ends of wire, bits of cut-off rebars, and similar items.

Japan is now in the last stages of this phase. The junkyard in Fuji is a classic in terms of the hypothesis presented here. Moreover, inventories begin to be a bit of a problem. In earlier stages, it was no trouble to dispose of almost anything immediately, but now junk dealers have to scramble around for markets. In India, the owner of one worn-out tire might be in pretty good shape, since he has a very liquid asset; in Fuji, he might have to collect a truckload or more before anyone will take them off his hands. Some kinds of junk, such as old bottles, may well be virtually unsalable for long periods.

The Business of Junk

It is at this stage that the junkyard comes into its own. In the earlier phases, virtually anyone laying hands on junk used it or sold it; now we have the age of the specialist, the firm that collects junk and sorts it out for all possible end uses. And the entrepreneurs working with junk may well be the shrewdest guys around as well. It takes considerable business sense to make money out of junk, and those interested in finding managerial and technical talent might well spend some time with these people. Since no one in his right mind ever pays any attention whatsoever to junk, such men are never known, studied, or even considered as a part of the power elite. But they may drive the biggest Cadillacs and Mercedes around.

Along with Japan, Ireland, Spain, and perhaps some of the East European countries are now in the piling-up stage. One can check this point by determining just how many junkyards per capita there are. Note also that at this stage the auto wrecking yard begins to become visible. It is only at this stage that countries have enough cars around long enough to generate any good supply of auto junk.

Junk and Morals

Outside observation of junkyards can also indicate something about a country's moral posture. In some places, the yards have high walls and twenty-four-hour guards. In others, like Japan, the gates are open, there are no guards, and almost anyone can wander around looking at stuff. When this is combined with piles of two by fours stacked on sidewalks, plywood panels leaning against walls, ready to use but unguarded, and rolls of copper wire casually stuck in corners of dark streets, it is clear that personal morality is pretty high. We should be so lucky in many other countries with the same per-capita incomes, but quite different social organization and ethics. Per-capita income in such situations ranges from $500 to about $1,200 per year. As income goes up, of course, physical consumption does too, so the number of junk dealers and the supply of trash also rises. This leads to more junkyards and much more trash. Perceptive observers can note the number of enterprises devoted to trash per capita—if it seems rather high, and if the junk is of the affluent type, then the country is well along the development path.

4. The age of affluence. As countries get richer, they of course pile up much more junk. Moreover, labor costs are high, and it begins to become economic to waste materials rather than labor. Once again junkyards change. Really useful stuff begins to appear, such as ends of copper pipes, four-foot lengths

of two-by-fours, ends of rebars, and coils of copper wire. The auto scene changes completely, since now it doesn't pay to take cars apart. Hulks contain all sorts of useful parts, and there are many more of them, scattered all over the place. Beer cans pile up, usually in all the wrong places, because no one bothers to pick them up. The country begins to get into the disposable container world, where things get used once and thrown away.

Influential people begin to talk about ecological problems and pollution, and campaigns to clean up things begin. Instead of being an asset, trash is now seen as a costly burden. And there appears to be so much of it!

Most of Western Europe is now in this stage, including all of the EEC countries and most of the EFTA nations. Since most of these countries also are growing rather rapidly, and becoming much more affluent, they are passing through this stage rather quickly.

In this stage, junkmen serve a less useful and perhaps even less profitable role. Things that in the previous stage were usable if they were taken apart and marketed right now become too cheap to bother with. Moreover, such countries typically have all sorts of lucrative opportunities for fun and profit for clever entrepreneurs, so why bother about junk? There is bigger game afoot. A possible problem in such cultures also is that while papa made his pile in junk in the last generation, in a level-3 situation, his heirs may not be so interested in these vulgar goings-on. This author has never heard a junkman admit that he was a junkman, and proud of it—in spite of the useful social purpose junkmen serve. It is curious that in all cultures the producer is widely respected and honored, while the fellow who moves the junk out so that more production can be accomplished is invariably of low status.

5. Total Affluence. In this culture, it sometimes appears as if the culture is drowned in its own garbage. Everything is disposable, including such high-value items as autos, and

the countryside is loaded with stuff that no one wants or can afford to pick up. Since people have long since forgotten the basic law that whatever goes in must come out, the pile up of disposables is tremendous. Ecological concern intensifies, yet somehow no one knows how to get rid of the junk. The point is reached where a reasonably intelligent man can literally live for nothing on the throwaways of the society. (One of the author's friends happily does so in Los Angeles. It's really quite easy.) In short, junk becomes a national problem, and presidents' wives start campaigns (which don't work too well) to clean up the country.

This is the American situation now, and quite a few European countries are rapidly getting to the same position. But trash can only be a real problem if you're so rich that it really doesn't matter too much. Materials that are valuable elsewhere, such as paper, are so plentiful in used form as to have almost no value at all.

Conclusion

Now we have a junk model, pregnant with implications. The problem is to put it to use, which is what the rest of this book will be about. Because this pile of used paper is published in the United States, and because most readers affluent enough to buy it are located there, we'll begin at the end of the junk cycle, by considering the period of total affluence. The United States is already there, but Canada is close behind, as is Sweden, Switzerland, and a few other wealthy countries. What we will be doing is figuring out not only how to read junk, but also how to use it to your advantage.

Since I try to be a university professor, I always like students to do some homework. In studying junk, this is very easy—as you read, just look around you, wherever you are. If you're in the United States or Canada, you can see more junk and garbage in a day than you can handle. You don't even have to handle it—just look at it, and for the first time, read what it says. And as you do this, you are on your way to real wisdom in the way the world operates.

2

Total Affluence: The United States

As countries become affluent, they get into serious junk problems. This is because no one really has thought about the key rule that what comes in, has to go out, somehow. In a country like the United States, which uses very large amounts of petroleum, coal, water, and everything else, the *out* part gets rather nasty, as air pollution rises, garbage gets very hard to dispose of, and waterways become moving sewers.

We suddenly seem drowned in our own effluents. Five or ten years ago, people wrote learned and perceptive books forecasting the long-term future without once referring to pollution problems. The sudden tide of crud has overwhelmed the republic, caused politicians to run for cover or spend millions cleaning up, and generally shifted future national priorities. What happened?

Actually, if anyone had bothered to look, the problem could have been forecast with considerable accuracy. But the United States, like everyone else, has a social hang-up which virtually guaranteed that the junk problem would come upon us without warning. Other countries, to date less swamped with junk, will also find to their dismay that the problem remains unanticipated and probably unsolvable, for the same reason.

The Social Status of Junkmen

The reason is that junkmen and garbage collectors have

absolutely no status in anyone's society. Hence they are ignored and despised. Who is the General Motors of American junk, if such a firm exists? Who cares? Those who pick up the garbage after us are probably semiliterate immigrants, or something; that they may be quite efficient businessmen and managers is irrelevant. The glorious, high-status people have always been the producers, those who shove still more stuff into the system for us to consume and enjoy. After we have done so, some low-caste character can pick it up, if it gets picked up at all.

Where junk dealers are mentioned, it is to identify their low status. *Sanford and Son,* the TV series, is about a couple of junkmen, but the stories never deal with their junk—the stuff is nothing more than a background prop for stories about the characters. As junkmen, they are definitionally at the bottom of the ladder. And no Women's Liberation group argues that junkmen should become junkpersons—this is one area that can safely be left to the males.

This social attitude is common because until quite recently all countries, including the United States, were scarcity economies. No one had enough of whatever the good life required, so anyone who could get more efficient at production was a hero. And, as we shall see later, poor countries with real scarcities rarely have garbage or junk problems. Whatever paltry goods are produced get totally consumed, and junkyards are few and far between.

So we make heroes of producers and despise junkmen. Capable citizens, correctly perceiving what the reward system is, go into production, invent new things, find out how to do old things more cheaply. In fact, by around 1965, they had succeeded to the point where most people could have all the goods they wanted, along with lots they didn't really want but had anyway. *Fortune* and *Business Week* run articles—illustrated profusely with pictures of the productive heroes—about their accomplishments. Through such propaganda we subtly encourage others to emulate these successful producers. And in business schools and elsewhere, we provide educational opportunities for new production heroes. But aside from a bit of sanitation engineering quietly taught in engineering colleges, no one is even able to find out how to be a successful junkman. I have looked at tens of thousands of business cases used in all the best schools, and as yet I haven't found one dealing with junk in any form.

The U.S. as a Trash Generating System

Now, when all that production finally comes out as junk, sewage, air pollutants, or garbage, we find that we don't quite know what to do. Up to now, some poor fellow took care of the minor problem, but now it is a major problem. And we are overwhelmed by stuff we don't want and can't handle.

In retrospect, it is easy to see how we got where we are, but it may be useful to sketch this process briefly, if for no other reason than to alert readers in other countries that are rapidly getting into the same tough shape. It goes about like this:

First, the United States has been getting steadily richer for a long time. Income per capita now stands at around $4,200 per year. This is another way of saying that labor costs have also gone up, since the way you get rich is to get productive. And as workers get more productive, they tend to get paid a lot more. One unexpected byproduct of this development is that anything that can't be made productive tends to get very expensive. These days Americans bus their own dishes in cafeterias, which increasingly replace the older restaurants where waiters served them personally. Waiters get paid too much, so to keep meal costs down, we now have all sorts of automated serving systems. We also use home electric washing machines instead of maids or even commercial laundries; we park our own cars, and fix them too, if we can; and we tend to do all sorts of things that used to be done by low-paid flunkies. Those low-paid flunkies are now highly paid technicians and workers in automated plants and offices.

This is important for junk, since so much of the junk and

garbage business is very labor-intensive. As wage rates go up, so do garbage-removal costs.

Second, energy costs tend to decline in a rich society. We learn a lot about transforming coal, oil, and water into usable energy to do our work for us. And we find that products that require much energy to make, such as steel and aluminum, become cheap and plentiful. Aluminum costs about as much per pound now as it did in 1930, although prices in general have more than tripled since that time. If something is cheap, it tends to get used in ever-increasing quantities.

In rich countries, we also learn a lot about chemistry, physics, and industrial engineering, so we find that lots of plastics and other new materials are available. It is hard to remember that only forty years ago there were only a few scarce, expensive plastics around. Antique collectors these days find that toys in 1935 were largely made of iron, lead, tin, brass, and steel. To really appreciate this point, consult a copy of the 1927 Sears Roebuck Catalog. Most clothes were wool or cotton, although a few minor rayon items were seen, and if you were stinking rich, you might buy a pair of real silk stockings. But synthetic fibers were virtually nonexistent. Forty years ago we were much more ecologically harmonious with our environment, using (in much smaller quantities per capita) only things that were a natural part of our world.

The Paper Revolution

Very importantly, paper gets much cheaper in rich countries. Papermaking is a very complex, large-scale industry these days, and companies know how to make it cheaply in enormous quantities. Hence we take for granted such things as 200-page newspapers, free shopping bags, and big cartons to put anything we want in. It was not always thus—once again, not too long ago such things were expensive and not very plentiful. You took your groceries out in your own shopping bag, not in the store's free one, and when you bought a big item like a washing machine (which were produced in small quantities), it didn't come in a huge cardboard carton. And your local paper was four or maybe eight pages, if indeed you could afford a paper at all.

The Good New Days

As we grow rich, more of us consume more. One notable change in the United States in the past thirty years is that the basic consumer-goods package is now available to most families. We find that from 80 to 98 percent of American households have a washing machine, a drier, several radios, a TV set, a car, and so on. And these items, contrary to what most people think, actually are getting steadily cheaper in real terms, not more expensive. You old-timers can remember your first black-and-white 1947 TV set, which was about as big and heavy as a house, and cost maybe $500. You may also remember nostalgically your 1939 Packard, which cost more in real terms than a modern Oldsmobile or Dodge, and which was a lot less comfortable and safe besides. You may remember that you bought your wife an iron mangle to iron with—now we pay a bit more for permanent-press shirts, and who irons anymore? Durable consumer-goods prices actually have come down a bit, and since real incomes have gone way up, lots more people have these things.

The household gadgets are not only cheaper, but they often are much more complex and hard to fix than before. And, of course, here we run into labor costs again. Something that was manufactured on some assembly line by a few machines may be virtually impossible to get apart and fix by hand, as the maintenance man has to do. Ralph Nader to the contrary, most household stuff these days is much more maintenance-free than it used to be, because smart businessmen, observing both that maintenance costs were rising fast and that customers screamed when things broke down, have designed their products to last longer. I know that no one believes this, since things do break down all the time, but you should have seen what was going on twenty-five years ago! It's a lot better now. But combine more difficult maintenance with very high labor costs, and, when things finally do break down, you get lots of new junk. In the old days we fixed; now we throw away.

You might ponder how pervasive this point is by considering how you would behave if your TV set stopped working and you discovered that it would cost you $4,000 for a new one. My guess is that you would call in the repairman, let him fix it, and sigh and pay his bill for $47.95. Now you toss the five-year set out with the garbage and go down to buy the latest, better model for $129.95.

What we have done consistently is to make capital cheap and labor dear. This means that we automate factories to make things cheaply, at the same time making it very expensive to fix the things we already have. No one did this with any malice aforethought—indeed, it seemed the humane way to go, since one nice effect was to raise most people's incomes. But the side effect is to generate lots of new junk that no one really thought much about.

Rags, Bottles, Cans, and Pileups

As we get rich, we also get more impatient. Time is one thing we cannot change, and we only have so much of it. Hence anything that saves time is used. These days we find that the cost in time of preparing and packing food is more than the raw food cost. In the old days, grandma bought her peas fresh at the grocery, took them home in her own shopping bag, and laboriously shelled them herself. The only garbage generated was the pea shells, which were often eaten by the hogs. Now we buy a nicely packed frozen peas package, where all the work is done, and we take them home double-wrapped.

We rip off all that cheap paper and toss it away, open the package and toss it away, and heat and serve. The result is time saved—and garbage generated. Multiply this by the millions of other sorts of prepackaged and prepared foods and other items, and we have an unexpected paper and plastic junk pile on our hands.

Moreover, producers found out in the mid-1950s that most people hate to mess with returnable cans and bottles, so, given lower glass and aluminum costs, higher incomes, and time to be saved, we entered the nonreturnable can and bottle era. Once again we had to figure out what to do with the junk. We still haven't. Remember also that messing around with a bunch of dirty old beer bottles is very labor-intensive, and wage rates keep on going up. Even if we try to return to the good old days, we will find that costs are way up, and even at 25 cents per returnable bottle, it may not be worthwhile. You can automate making and filling the new bottles, but it is very tough to automate the return of old ones from a million scattered points.

Rich Man, Poor Man, and Still More Junk

Marketing men have long since discovered that the United States has a curious income-distribution problem. About a third of the families can afford almost anything they want any time they want; another third can sometimes afford most of what they want, but not quite all, and often they must buy used items; and the lower income third can't afford many necessary items, particularly consumer durables. These thirds are relatively stable; if you are now in the top third, you are likely to be there ten years from now. You will be richer absolutely, since incomes go up all the time, and the people in the middle third also will be better off, but your relative position is not likely to change all that much.

The problem with durables, from the manufacturers' point

of view, is that they last too long. How can you get that top third to buy a TV set or a car if they already have a good one? The easy answer is forced obsolescence. Get the consumer to believe that whatever he has is out of date, lacks useful features, is designed wrong, or whatever, and he just may buy a new one. This is nice, because he can trade in the old item, which the dealer can sell used to the fellow in the middle or bottom third. When it finally wears out, it goes to the junk pile.

Junk Generators

Our federal government, in its zeal to protect consumers, also helps generate consumer junk. Each year higher safety standards go into effect, and they are quite likely to continue to be tougher. The various agencies publicize dangerous items; conscientious consumers get rid of them by tossing them into the junk pile, and buy new, safer ones. Since all these agencies are managed by persons in the top third of the income-distribution pattern, they tend unconsciously to adapt the attitudes of this group. Something, say a toy or electrical appliance, is seen, correctly, as unsafe under certain conditions. What more logical thing to do than to throw it out and get a new one? The idea that the other two-thirds of the population cannot afford to do this never occurred to them. Top-third Americans have long since forgotten that something is always better than nothing, because they always have had lots of things.

Actually the federal government is one of the biggest junk producers in the world. With federal budgets going above $250 billion each year, officials can buy a lot of junk. Much of the bill is for labor, but government departments do buy incredible amounts of hardware, and sooner or later it all gets tossed out, rejected, and sold as government surplus.

The biggest junk generator of all is the military, since no one really cares what things cost, nor whether they ever get used. A citizen may hang on to a defective hair drier for a while, but if a military system turns sour, the whole thing gets junked. And you can find thousands of square miles of military junk like warplanes, out in the desert someplace. The stuff gets sold for a few cents a pound, and gradually works back into the system. You can find lots of it in war-surplus stores and electronics-parts catalogs, like microswitches that cost new $14.92, now going, this week only, for 49 cents.

I always wanted to buy a Norden bombsight (new cost, $5,000; surplus price, $49.95), just to see what was in the thing, but I never got around to it. Others, more fortunate, did buy new $10,000 Allison aircraft engines for $50, and one Hollywood stunt pilot once owned something over 5,000 surplus military aircraft, which may have made him the biggest used aircraft junk salesman of all time. He made money on the deal: it turned out that all those planes had high-test airplane gas in them, and by the time he drained the tanks and sold the gas, he had paid for the whole fleet. This was just after World War II, but even now the military generates very high-quality junk in prodigious quantities.

This whole system is fun for lots of people, and it also can be very profitable, but it also generates lots of new junk. For minor items, like hair driers and toasters, the old one just gets tossed out. For big things like cars, it means that more new cars are produced than are really necessary just for transportation reasons. The old cars, which could run lots longer if they were worth fixing, get junked; the new ones generally go to the upper third in income terms; while the used cars find their way to middle- and lower-third families.

The Case of the Bewildered Repair Man

Just to make the problem more junky, being new means being different, and being different means that your friendly repairman had better be a very smart fellow. Whatever he did to fix the carburetor of your 1970 Buick is not going to be what he will have to do to fix it on your 1974 model; moreover, he will need two sets of expensive parts instead of one in his inventory. So your maintenance costs go up, and you read the ads about the new models, and you and your wife decide oh, well, let's go get a new one and let someone else worry about all that expense.

Add this thinking to the fact that now over 70 million families have 85 million autos, and that a third of them can afford to buy almost any time, and you really have some junk—like maybe 6 or 7 million cars alone each year.

Out of Sight, Out of Mind

The steady trend to urbanization hasn't helped much either, particularly when the poor are in the inner city, while the rich are out in the suburbs. The rich pass consumer durables used to the poor, in whose hands they break down and get abandoned. Remember, the poor often do not know how to perform simple maintenance (which is one reason that they are poor), so even pretty good used items may get tossed aside. And then the junk is all in one place, instead of being scattered around the countryside. Few people care much about invisible junk—I know about a pile of 1930 cars on a farm behind a hill, which have been sitting there for over thirty years, out of sight, out of mind. But people care a lot about huge piles of rusting, visible scrap, broken bottles in the streets, or abandoned car hulks in their block. Not only do we have much more junk, but it now is much more visible to more people. So we worry a lot more about the problem than we used to. Farmers used to toss their old cans, bottles, and other crud into the nearest ravine. They didn't use much, so it took years to get a pile big enough to bother anyone, and no one but the farmer and friends ever saw the stuff. If he was a neat farmer, he could bury the pile every ten years or so also. But there aren't many convenient ravines in suburbs or cities, and people are so close together that the pile would grow too big too soon, so we need to organize trash and garbage collection systems to get the job done.

Minimum-wage legislation and uplifting poverty programs haven't helped much either. To see this point clearly, you might try arguing with a group of our present poor that they should be content with being garbagemen, even at good wages. No, they want the right status things, like being scientists, producers, computer programers, teachers, social workers, and all the rest. Garbage and junk is exactly what they are trying to get away from, and of course no nice middle-class suburbanite in his right mind is going to suggest to his children that they become junkmen.

Another key point is that the United States is rapidly becoming the first country in the world without a real lower class. Even the now exploited minorities, like lots of other earlier exploited minorities, are slowly getting into the mainstream. And those who can't quite make it will be helped by public welfare programs. Try suggesting that welfare mothers be forced to pick up the trash in their neighborhoods to see how far we have come along this particular road. Such work is degrading, unfair, cruel, etc.—which of course means that the junk will either be picked up by very highly paid fellows the city probably cannot afford or, what is more likely, it won't get picked up at all. In a world where everyone has enough status so as not to do the dirty work, who does it? Historically, we brought in some other lower-class immigrants to do it for us, but we haven't done this for fifty years now, and the old lower class is dying out.

Finding Good Junk

Another problem that stems from our residential class locations is that good junk can be found in upper-middle-class neighborhoods, while the people who might need it or be able to use it cannot get there to pick it up. A neighbor once offered to give me a running car, just to get it out of his way, but he was not willing to find some poor man who needed it in the distant ghetto to give it to. One can also find working washing machines, TV sets, loads of usable clothes, and much more in the wealthier neighborhoods, but few poor are around to get it.

In the old days, maids often took such stuff back home, but who has a maid these days? Once again, our own affluence, humanitarian instincts to protect lower-class working people, and the resulting locational isolation of classes has worked together to compound the junk problem.

Tax Avoidance and Junk

Just to make the problem worse, most Americans dislike paying high taxes, and as suggested above, hiring people to move garbage and junk these days costs lots of money. Often private firms do the job, but typically they work under contract for local governments, who get their money from the taxpayers. And when taxpayers rebel, as they often do, garbage collection suffers, along with schools and other public activities. It's very hard to get emotionally involved in garbage or junk, unless it's on your front lawn. Since most junk these days is really not quite that visible to more affluent taxpayers, why bother to support the city's plea for another ten cents on the tax rate to improve garbage collection, or finance special junk pickup work?

And to top the whole problem off, most Americans, all protestations to the contrary, are still a bunch of uptight Calvinists about dirt, garbage, and junk. To the vast majority, cleanliness is next to Godliness. Hence it is quite easy to get many Americans upset either by having some nice dirty junk someplace where they can see it, or by showing pictures and TV shots of great piles of messy offal, old cars, overworked sewers, or other manifestations of the "going-out" part of our postindustrial world.

The Great American Junk Machine

When all the factors are put together: the low status of junkmen; the focus on the producer as hero; the suburban explosion and the collapse of the inner city; the rapid and pressured obsolescence of consumer durables; high maintenance cost; cheap paper, aluminum, plastics, and all the rest; and the high labor costs, it looks as though we couldn't have done a better job in creating a junk problem if we had sat down and tried to figure out how to do it. Our whole culture is, by accident, beautifully designed to generate huge masses of disagreeable junk and garbage, to say nothing of air pollution and other nasty effluents. So, we perceive crisis, and wonder how to reform the situation. It can be done, but only if we spend a lot of time and energy turning around some very deeply held American beliefs. We'll get back to this point shortly, but first let's do some junk reading.

Reading Junk

Just as anthropologists can infer the nature of any ancient culture by looking at broken pots and other junk (indeed, the real archeological gold mine is an old dump), we can read lots of things about the American culture from just studying its junk. As we suggested above, American junk is not very monolithic—each neighborhood and byway tends to have its own particular idiosyncrasies. Anyone can play this game, and if you learn how to do it, you can apply the results to all sorts of useful things that need doing, including raising your own standard of living.

The really high-quality junk to be found is in upper-middle-class suburbs. Here families have more money than they really need, in terms of obtaining mere food, clothing, and shelter, and they tend to buy things. Boy, do they buy things! This is the paradise of the supersalesman with his consumer gimmicks and gadgets, and his brand-new models. And since modern suburban homes really don't have much storage space, when the new comes in, the old goes out. Incidentally, single-family homes are much better for junk than apartments, because there is *no* extra space in most apartments, so the stuff goes out more quickly. Moreover, the apartment superintendent may have a junkman friend who takes all the goodies off his hands for a small fee. Remember that while we are trending to an absence of an underclass, such characters still are around, and in any good upper-income apartment district there will be found a few junk professionals from the other side of town who make a living off the discards.

But suburban single-family homes don't have supervisors, but rather harassed housewives who, finally, scream out to the kids, "That junk has got to go, NOW!" So, in the alley out back, one can find real goodies, like TV sets that maybe even work (or perhaps need only a tube or two), old bicycles with flat tires, that electric carving knife that just never did work too well, piles of slightly thin blankets, old children's clothes, and much, much more. You won't find old cars, since most upper-income suburbanites trade in their junkers long before they stop running, and in the rare cases where they do have one, they can afford to pay a junkie to take them away.

You also find tons of toys and games since most toys break rather easily under the hard usage they get, or, if they don't, the kids get a bit older and lose interest. Such neighborhoods offer vast possibilities for living for nothing, which we will get into in Chapter 4 in some detail. They also let any observer know what kinds of people he is dealing with. Note that as one works down the social and income scale, the quality of junk begins to deteriorate ever so slightly. The TV sets are still there, but they are twenty-year-old monsters, instead of five-year-old transistorized models. The toasters are likely to be more beat up and older, as are the toys, which are more broken than they are in more affluent surroundings.

Many cities have junk pickup days, when the citizens are encouraged to pile up such stuff in the alley or out front. Salesmen are advised to spend some time walking around on those days, since you can tell more about the kind of families in the neighborhood by seeing what they throw away than you can almost any other easy, public way. Consider a fellow trying to sell health insurance. In wandering around the neighborhood, he will note that some citizens take a very casual, devil-may-care attitude toward life. It's all there in their junk—he sees almost new toys, only slightly damaged appliances, suitcases that are new, but with the hasp torn off, empty bottles of good scotch, and similar items. Next door, not only is there very little junk, but it really is junk—old medicine bottles, ancient filing indexes, eight-year-old magazines, and similar dreary items. Clearly this family is frugal, cautious, and careful. When he makes his sales pitch later, he will know about what to do and how to proceed. Any serious salesman who bothers to read such junk is likely to be a lot more successful than the one who just sees a good neighborhood, concludes that everyone is rich, and flies blind into all sorts of family situations.

Upper-middle-class garbage, if one likes or can stand looking into the closed cans, is very high quality also. Lots of useful food gets thrown away, and no one is around (usually not even animals) to enjoy it. Cans and bottles abound. The pets in such households get their special dog or cat food, not family leftovers. And as we move down the income scale, we find deterioration. The pets eat the scraps, and there also is much less paper, since the bags are used for lunches, and housewives still often buy vegetables raw and unwrapped, rather than precut, packed, and frozen. A family has only one garbage can, not three or four. You can tell a lot about a family by seeing what kinds of reading materials get tossed out too—clearly a big pile of old *New Yorker* and *Harper's* magazines suggests a somewhat different type of family than a smaller pile of *True Confessions* and *Police Gazettes*. And the man who discards huge piles of *The New York Times* is a very different sort of man than the one who puts out a thinner stack of *The Daily News*. Since the affluent tend to read much more than the less well off, we typically find much larger paper trails around the upper-middle-class neighborhoods than we do around more modest dwellings. Moreover, the middle classes use more paper over again, for lunches, wrappings, and cleaning, than the upper income groups do.

Car Twinning

As we move down the income and class scale, junk tends to deteriorate and change. Somewhere in the blue-collar neighborhoods, the junk cars begin to pile up. Often, if you pay attention long enough, you note the twin-car phenomena. A fellow will have a 1962 Chevy out back (or even in the front yard, particularly in remote locations), with the wheels off and miscellaneous parts missing. When you come after work and find him home, you will discover that his car that runs is also a '62 Chevy. What happens is that he has found (or bought for a few dollars) a twin to the car he runs, and it becomes his parts supply. This is very logical—you may pay as much as $50 for a new carburetor or alternator for your '62, but you can buy the whole car for ten bucks—or maybe get one given to you if you haul it away. The resulting junk pile is messy, but it is a very efficient way for middle- and lower-income families to keep their car running very cheaply.

Why does that carburetor cost $50? Well, you see there are so many model changes and different types that the parts house has to have huge inventories, and these cost a lot to maintain. Moreover, if the parts man makes a mistake, he eventually has to write off that $50 item because it didn't sell. . .

Total Affluence: The United States / 47

Of course, in the end both cars die, and the bones lie around for years, visible and nasty. I've seen as many as seven sets of twin cars in working-class single homes, where there was enough space to keep them (out in the country, but near town), and only one set actually in operation. Remember also that working-class people tend to help each other out informally. Joe's two '62 Chevies may not be running, but Sam, over in the paint shop, has his '63, and lots of parts may fit. So he talks to Joe, drops around one sunny Saturday, and borrows a starter motor, the voltage regulator, and a few other bits and pieces. After ten years of this, the twin cars are reduced gradually to totally stripped hulks. Then the county or city has a cleanup campaign, or an itinerant wrecker that shreds old hulks comes through, and the cars are finally banished from the landscape. In the rare cases where the hulks last over twenty-five years, an antique auto fan may buy what's left and haul it away. But in any case the process takes a long, long time.

Mass piling up of cars is rare in central cities, but in any neighborhood where garage space is available, or alleys with odd parking spaces exist, an incredible number of old cars are still around. My survey of my own middle-class (ranging from upper-middle professional to lower-middle working-class single-family and low-rise small apartments) neighborhood recently turned up such things as a pair of 1955 Nash Metropolitans; a 1949 Buick coupe; a 1948 Olds sedan; twin 1939 Plymouths; quite a few early 1960s cars in all sizes and shapes; and much more. This is in a city that has very strict rules about leaving cars on public streets, but none of these were on public streets, or, in most cases, even visible from any street. All I did was just walk the alleys behind the houses (which are public property) and observe the garages and nooks and crannies behind the houses. Some of the cars are in private driveways, which is legal, as long as you don't have more than two without licenses visible. None of these cars ran; they were just sitting, unlicensed.

Note, though, that you typically do not find such things in upper-middle-class neighborhoods. And, as we suggested earlier, the quality of the junk tends to deteriorate markedly as incomes go down. In our town, which has a lot of poor but quite bright students, the trash days are disappointing, since the young people go out early and skim the cream off the stuff before the junk collectors get to it. Lots of young people without much money have learned all about junk. They won't talk about it much though—why spoil a good thing? One young man I have watched with admiration for a few years, who lives on junk in our small city, refuses to even talk to me about it. He is afraid, quite properly, that I may steal some of his secrets.

Poverty Problems and Poor Junk

As one gets to the lower-income parts of any American city, junk deteriorates still further. Broken bottles, old chunks of rotted wood, and similar crud abound, but really good junk is hard to find. One major reason is that poor people have less to throw away; another is that when things finally do get discarded, they really are beyond repair. One rarely finds any cloth in poor neighborhoods, as one example—any piece of clothing, no matter how small or torn, can be used for something. The upper-income person buys special, absorbent paper towels to wipe things with; the poor person uses an old rag, which used to be a child's shirt.

But in *really* poor neighborhoods, if you dare go there, the quality of junk goes up again. This is the "nothing gets used" part of the world—really poor people are typically not well educated, and they tend to throw away almost good items, like cars, which they do not know how to fix. Usually such poor neighborhoods are also very congested, so even if a person wanted a twin car, he could not find the space to put it.

This neighborhood also is the end of the line for much hot merchandise. If a TV set didn't cost anything to begin with, it is likely to be tossed out rather casually when some

Total Affluence: The United States / 49

minor thing happens to it. A person who stole one set can go back and steal another.

But the really poor neighborhood is not good hunting grounds for the serious junk collector. The reason lies in the high population density of the area, plus the large number of persons hustling to make a buck. As soon as good junk is discarded, it gets picked up for resale. I once watched what happened to an apparently abandoned car in such a neighborhood. It sat untouched for two days, a battered eight-year-old four-door sedan. Then, on the third day, the tires and wheels vanished, to be followed the same day by the battery and headlights. The next day all the windows were broken out, and that same day the engines, radiator, and all accessories were neatly taken out in about an hour by two fellows with a beat-up, unmarked wrecker. The hood disappeared on the fifth day, along with the bumpers and the right front door. The steering assembly went the next day. The taillights lasted another day, surprisingly (they usually go first). Then, except for the kids having fun throwing rocks and banging the thing up, it sat for almost six weeks, until the city wrecker came to haul the rest of the car to the bone yard. A rather new, but unworking TV set tossed out on the street in the same neighborhood lasted less than forty minutes.

It depends on where you are as to what happens. City sanitation departments in bigger cities pick up everything from complete cars that have sat around for a while (often abandoned by a stranger in some middle-class place like a suburban shopping-center parking lot) to hulks with nothing to them. But it is rare to find a complete anything—car, TV set, or any consumer durable, as junk in really poor neighborhoods.

Industrial junk

Industrial junk, of which there is lots, rarely gets talked about much, except by professionals. This is because most of us rarely see it casually. Industrial plants are now located way out someplace, and the junk they generate tends to get disposed of quietly and without fuss. Only when a plant pollutes, as when its smokestacks pour out noxious black gunk, do we get much upset. We also get very uptight these days about water pollution too.

Industrial junk can be read just like any other kind, though. As you might suspect in a country with very high labor costs and relatively cheap materials and energy, our industrial junk reflects the intensive use of labor. We will not worry much about generating more scrap if it leads to labor-saving. Since industrialists are quite logical, cost-conscious people, they tend to be very rational about how they generate junk. It is common to find such items as the cut-off ends of brass rods, large amounts of scrap iron, piles of almost new lumber, and similar nice things around industrial plants. But this material is easy to handle, and the professional junkmen know that it will emerge in just the right mix and at some given rate. Hence it is easy to make some long-term contract with various junkmen to pick up the stuff when necessary. The company can easily arrange to have different kinds of junk come out in neat piles, all sorted and ready for pickup, which is exactly what households cannot do. If you want to pick up household junk for ten thousand homes, you have to pay someone two to five dollars an hour to do the sorting for you. But the industrial plant merely has to instruct its workers to pile the brass shavings in that bin, the cast iron here, and the wood over there. When the junkie* comes, it's ready to use.

Junk Users

In spite of the fact that we generate an oversupply of junk,

---

*"Junkie" is a debased word these days, having been taken over by not very nice producers. But its meaning here is the original.

it is true that we have many professional junk users and handlers. This is a market that rarely gets talked about, for the status reasons we noted earlier. But some of the wealthiest men in the United States made their money handling junk, and fortunes can still be made with the stuff. If the junk dealer can get the stuff in reasonable quantities, in well-sorted lots, he can usually figure out how to make some money from the deal.

Since no one gives a damn about junkies, there is of course no government regulation, and no data about the industry. The market is highly organized along the lines suggested by Adam Smith—free-for-all enterprise, and damn the fool who tries to do something humane. One also suspects that junkies are quiet about their activities, since not only is the market free, which means that anyone who really knows what is going on can get in on a good thing, but also so much of it is done in cash and beyond the review of government. That can mean nice tax-evasion possibilities. One rarely hears of a serious government accountant messing through huge piles of greasy cast iron, looking for new tax possibilities—and if it doesn't happen, what is to stop a sharp operator from conveniently failing to report certain revenues?

Trying to regulate or control our junk markets would be rather difficult, since anyone with an old pickup truck and a place to put the stuff can play. There is nothing stopping you from going out tomorrow, buying or stealing some trash, and selling it to anyone who will buy it for profit. If you can do it, so can thousands of others, and they do. If our government set up an office of junk control, it would be years before they even found out who was in the business, let alone were able to do anything much about it.

Junk for Profit

Some junk is extremely profitable. Copper, as one example, can be remelted and reused easily, and copper these days sells for thirty to sixty cents per pound. Aluminum, steel (if sorted properly), tin, zinc, lead (which is why those car batteries go so fast), and other metals are easily reusable. Junk dealers in this field are often large, reputable firms, with the ability to handle thousands of tons of materials per year. There also are many organized junk markets, some of which even get reported in the *Wall Street Journal* (as I'm writing this, I note that number 1 melting steel, whatever that is, is selling for $46 per ton, FOB Chicago). A complex supply/demand system is organized to flow usable scrap from generator to reuser.

But only industrial firms are interested in tons of copper or brass. The other kind of consumer-generated junk flows into quite different channels. Historically, the very poor used it, and some still do—you can find used clothing stores in poor neighborhoods in many cities, and occasionally you can even find older men laboriously going around collecting various consumer junk for resale. The Salvation Army and Goodwill are big junk users; inoperable TV sets and refrigerators are recycled by handicapped employees, and sold in the organizations' own outlets. Other charitable organizations also collect reusable materials and do the same. These are very good consumer deals, incidentally—I recall fondly the Goodwill Department Store in Santa Monica, where we once bought an excellent recycled refrigerator for $30. When it stopped a week later, a Goodwill repairman came and fixed it, and it ran for ten years after that. And in that store the quality of the clothes, which came from very affluent parts of Greater Los Angeles, was excellent.

One encouraging trend these days is that upper-middle-income young people are beginning to buy their clothes in such stores. They take the position that poverty is nice, particularly if it is pseudo-poverty, and my college classes these days are full of old army coats, jeans once owned by farmers, and other used items. Army surplus stores, which these days sell a lot more than army surplus, also provide nice outlets for excess inventories and slightly used items.

Doing Something About Junk

For Americans used to a more regulated life, junk seems to follow weird economic laws all its own. Actually, the market is merely reacting the way any competitive market does—if there is a buck to be made in junk, someone will be in there making it, if he or she knows about the opportunity. This point suggests a series of very powerful policies we could adopt if we really got serious about using junk more efficiently than we do now. What has happened, as we suggested earlier, is that the price of new things is too low, while the supply of old things is too great. The former condition motivates people to buy before the old thing is used up, while the latter keeps the price of junk down, thereby lowering profitability for the junk seller.

Hence, if we wanted to get rid of more junk, we could just structure taxes so that new stuff costs more, which would lead to old stuff being kept longer. As this happened, the price of the old would also rise, making it worthwhile to salvage. We could also do some interesting things by linking up salvage to initial production. We will get at this in detail in Chapter 4 in talking about cars, but what is involved is making it capital-intensive to salvage junk, rather than labor-intensive, as it now is. That is, any junk we generate should be very easily separable into its original elements for recycling instead of being hopelessly mixed up so that it costs too much to separate.

Someone is likely to pick up an old TV set, even in the United States, but in the case of most poor junk, like cans and bottles, the economics of the problem are totally in favor of just piling it up. Bottles, cans, and plastics containers flow out to 80 million households, which use the contents and then toss the containers in some gooey garbage. To get them back for recycling, someone has to develop a sophisticated screening system, and financially pressed cities would have to pay for it. And after the bottles, steel cans, aluminum cans, and various types of plastics are neatly sorted for recycling, they have to be crushed, packed, and shipped, often hundreds of miles, to a plant that can use the material. You can get around $200 per ton for mashed aluminum cans; $10 per ton for sorted bottles; and around $8 per ton for old newspapers—which just isn't worth it in most places. So the junk piles up in landfills, where it at least serves a purpose, and in other places where it doesn't. Of some 36 billion bottles consumed annually in the United States, only about a billion are recycled; of around 64 billion cans, only 1.5 billion are sorted and scrapped.

It is fun and good citizenship to get the Boy Scouts or the local chamber of commerce to set up recycling bins and urge citizens to turn in sorted materials, but such efforts have one major defect—they run uphill economically. That is, it takes time and trouble to do the sorting and returning, and after the initial enthusiasm, people normally have other things to do. Moreover, schemes such as this assume that some market exists for the materials, which in many localities is not true. Few companies can handle old paper, and efforts to collect more of it can just lead to huge piles of paper in some other location than where they would have wound up if no one cared. To recycle materials properly, one has to design the entire system, and so far this is not even being thought about very much.

People also suboptimize terribly in their recycling thinking. I've seen matrons take four bottles five miles in their V-8 station wagons to the recyling center; one wonders if the resulting air pollution and using up of the car and fuel really was worth it for the $0.000001 the bottles were worth as recycled raw materials.

Economists call these factors externalities. The total costs are not being borne by any individual. The producer of the can, the processor who fills it, and the consumer who uses it all pay for direct manufacturing costs, but no one pays for pollution costs. If I toss a beer can on the side of the road, I don't pay, and it doesn't really matter—but if a billion beer cans get tossed out, then it does cost society a lot. Since I can evade my share of the cost, why bother?

The solution to this problem is to restructure prices, through taxes, so that those benefiting pay. Here, a tax on cans at the source could be paid by anyone using the cans. The funds received would be used to clean up the mess and recycle the materials.

Already there are a few steps in this direction. The state of Oregon has banned nonreturnable containers, which is certain to raise costs for all sorts of soft and hard drinks. But the effect will be to recycle lots of junk and keep it out of streets and parks. Sweden has a junk tax on new cars—the idea is that any new car will be a junker someday, so the new-car buyer pays for the junking when he gets his new machine. Proceeds from the tax are used to pay for getting rid of the old cars. It would be easy enough to expand this idea to TV sets, washing machines, and other hard-to-junk items, thus raising new prices and encouraging the proper disposal of the old. We could go even further, and give back to the person who brings in the old item the fee once paid for the new one. If anyone received ten dollars for an old washing machine taken to a proper disposal point, all we would have to worry about would be theft of machines in use, not piles of junked ones lying around.

Something like this will eventually happen, but probably not soon. The reason is simple. We are still a nation of producers, not junkers, and if we slow down sales of new stuff, what will we do with all the unemployed? We haven't figured that out yet. So far, in spite of lots of talk about pollution, garbage, and junk, there is not too much evidence that we are really serious about doing much about it, except talk, and maybe worry a bit.

But in the end, as more and more people are beginning to realize, we do live on a finite planet, and we really don't need most of the stuff we now produce at so much cost of real and scarce resources. We don't need a new car every five years, or even every twenty-five years (my 1931 Model A still runs just fine, and it is suitable for most local purposes). We don't need the latest washing machine, or hair drier, or whatever—the old one, or maybe even none at all, will do just fine. And anything durable can be fixed and kept going, if it is worthwhile. Right now it is not, so we pile up junk. In the age of total affluence, we find, to our surprise, that everything that went in really does come back out, and we are drowning in our own garbage. Someday, when we get a bit more mature and perhaps more organized, we should be able to do better than that. Indeed, given our limited resources, the United States may in the end be the first country to come full circle, and wind up in the everything-gets-used category once again. Already some kinds of good junk is beginning to go this route—the problem is to figure out how to get more of the stuff handled in the same way.

The United States is way ahead of most countries on this total affluence route, but as other countries follow our economic lead, we can forecast gloomily that they will end up in the same place. Already Canada, Sweden, and Switzerland are right on our heels, and the rest of Western Europe, Japan, and Australia are right behind. One expects that as these nations reach previously unrealized levels of affluence, everything said here will apply to them too, until they also figure out how to manage the *out* part of the economic cycle. Hopefully, by the time they get to this stage, we will have figured out how to handle the junk, and other countries can benefit from our experience.

3

Junking the System

So you've finally had it, and you decide to get out of the rat race. This business of living in the split-level suburban trap is just too much, and that subway commute every day is driving you up the wall. Your job, while admittedly well paying, is a drag, and all you see every day is the running idiots who don't know where they're running. As you squish through the dirty snow on the sidewalk, your mind turns to subtropic climates, warm soft nights, and freedom from it all.

Or maybe you're not dropping out, because you've never been in. Lots of young people these days view the rat race with horror, and they are determined not to get suckered into it. So you are staying away from everything modern America represents.

Now, these are noble thoughts, and it would be nice if lots more people really did drop out instead of just talking about it all the time. And those who have dropped out have a bad habit of reentering, because they hadn't read this book yet. I see ex-commune types often, and when I ask them pointed questions about their experiences, they shuffle a bit, look at the floor, and mumble about incompatible comrades, and slobs who wouldn't do the work. But what really happened is that they dropped out without thinking through the problem carefully enough.

The difficulty is that it takes more guts than most of us have to really drop out. Going all the way to some tropic island sounds idyllic—until you get your first abscessed tooth. It's a great deal—until you discover that someone, like you, has to do all the dirty work, such as plow a field with a stick, and then get a crop that's so small that you're going to starve to death before the next harvest. It's even disconcerting to discover that *someone* has to carry water eighty yards from the well. The reason there *is* a mainstream America, with the split-level houses, electric appliances and all the rest, is that most people enthusiastically avoid really dirty work, and they like things like the pill, antibiotics, indoor plumbing (especially in winter, and always when you smell the outhouse), and tractors to do the hard work instead of human muscle.

But this sad fact about drop-out realities can be easily avoided if you learn more about junk. There is a fun middle ground, as more and more people are discovering, which is not only feasible but reasonably comfortable, and cheap besides. You drop out, but you stay in far enough to get most of the benefits of modern civilization without paying much of the costs. It can be done right now in the United States for a few thousand a year, and if your Aunt Minnie happened to leave you a small legacy, or if you've managed to save enough to get that much a year, you can have your cake and eat it too. You can even earn enough from junk to get by. What is involved is nothing more than becoming a creative junkman.

Being a creative junkman involves a few tricky and difficult things, but after a few years, you will be so far ahead of the pack that it is well worth it. Moreover, there is real satisfaction in doing it. I have noticed that the guys and girls who are always talking about dropping out, who always seem to be most violent about what's going on in the mainstream, are unfortunates who can't *do* anything. They can think very well, and they can write and do all sorts of intellectual things, but they have never had the pleasure of fixing up something with their own hands. And that is what junk is all about. People in the affluent society throw things away because they don't work. If you're going to live off the droppings, you have to make them work again. And in so doing, you can learn more about yourself, and gain satisfaction in doing something real. Then the whole drop-out experience can be a great personal pleasure.

Most American dropouts also tend to be upper-middle-class types (UMCs), and for them the first rule must be to get over the typical UMC hangup that thinking is better than doing. Plumbers, mechanics, carpenters, and other Archie Bunker types *do* things; nice people think. There are a few exceptions, like fellows who build boats or do excellent cabinetwork in their basements, but these guys typically are seen as a bit odd by their UMC fellows. Moreover, they may be mindless, avaricious middle Americans, too, since it costs money to get good power tools. But if you can accept the idea that nice people like you are perfectly able to do some skilled manual labor, you can live off junk, avoid the mainstream, be comfortable, and have a ball laughing at all the poor slobs who haven't figured it out yet.

How To Do It

Step one is simple. Get a car or pickup truck. Mobility is very necessary in the creative junk business, as is some storage space. Enough people already are used to hauling things around so that the price of a good used pickup or van is quite high, but here is where your education can begin. Remember, you can't do everything in a day, and you should plan to spend a year or more finding out how to succeed in this drop-out business. You never succeeded in anything else without thinking the thing through and getting trained for it, so why expect to be a successful dropout in a day? It takes real skills.

A good way to begin your education is to go to the new

auto agencies and ask them about really beat-up trade-ins. The agencies that guarantee their used cars are the best, since they will wholesale any dubious trade-in they get, and even the most posh agencies will get some really beat-up vehicles traded in from time to time. Cultivate a salesman and let him know what you want—which is a barely running, eight- to ten-year-old vehicle. For a hundred dollars or less, you've started to drop out.

But don't buy the first junker some slicker passes on to you. If you haven't done much of this before, any good salesman will see you coming four blocks away, and he'll be ready for you. You can wind up with the worst, hard-to-fix car that's around. Do some research first. One kind of research is where to buy a good car cheap; another is how to check it out before you buy. A book like Ray Stapley's *The Car Owner's Handbook* (New York: Doubleday & Company, 1971) has a chapter in it about how to really look at a used car—buy the book (or better yet, read it in your local library), and make some notes.

It is interesting that such books are never found in Goodwill stores, or even very often in used bookstores. You can find novels by the ton; college textbooks on any subject by the thousands; and works of philosophy without number. But the manual on Dodge pickups you have to buy, or you try to find at your local library. And if you do manage to find it in, you will discover that it is dog-eared, greasy, and well thumbed. These books are decidedly not junk, since the trail you are beginning to follow is well blazed by clever fellows who know already exactly what to do. But relax—some million cars get abandoned every year in the United States, and 5 million more are scrapped by auto wreckers. Somewhere in that mess, you can find what you need.

Besides the local auto dealers, do some talking and looking around. College towns, where people move in and out often, are excellent places to find good, old machinery. Often a student or professor is going overseas and has to sell quick. Both are noted for a tendency to keep ancient cars running. Military bases are another excellent source, since servicemen frequently have to sell out quickly. Older suburbs sometimes have those fabled little old ladies who have 25-year-old cars with 18,000 miles on them (I've got my eye on one of those, a '46 Chevy, just around the corner from where I live; in fact—maybe next year). Actually, the best cars to buy are around seven to ten years old—these are old enough to be worth very little, but young enough so that junk yards and parts houses still have cheap things that will fit.

Try newspaper want ads. In some places and with some dealers, if you have a ten-year-old car, even in running shape, the dealer will give you a hundred bucks off your new one and tell you to keep the old one—it's too hard to mess around with and sell for a profit. Such owners may try to unload on their own through a newspaper ad.

While you're waiting and looking for just the right car, get down to your local public library and do some homework. Remember, you're going to have to fix the thing, no matter what happens. Any library in any town over 5000 people will have lots of good information about such things, since plenty of people in town do their own car-maintenance work, and there is plenty of demand for such material. This sort of stuff is even reaching the UMCs, as the book *Keeping Your Volkswagen Alive* suggests. You will find at first that everything seems very mysterious, but after all, you're a logical person, and a car or truck is nothing but logical. Things like *Dyke's Motor Manual* abound, as well as all sorts of materials written for the laymen about the mysteries of fuel, electric, exhaust, and other automotive subsystems. And, if you're really serious, you can often find night-school classes in auto repair, ranging from one my aunt took for total duffers to highly advanced technical training. So start getting educated. Remember, when you drop out, you won't have the hundred bucks to pay the repairman every time some little gismo breaks, and it's a long walk to anywhere in the United States these days.

Get educated about parts too. Visit your nearest auto junkyard (more about these in the next chapter), and find out what kinds of parts are stocked and how much they cost. Get (for fifty cents from most newsstands) a copy of J.C. Whitney's latest parts catalog and scan the pages, seeing what costs what, and what is available (they go back to 1940 on most cars, so don't worry about being too out-of-date). Visit your local auto-parts stores and see what kinds of things are sold, and for what prices.

Tools for Serious Dropouts

Buy some tools, after some research on what to buy. Good tools always cost money, and they always save or earn money, so don't scrimp here. Buy this sort of stuff before you drop out, so you can afford the best. For under $200 you can equip yourself to dismantle or diagnose any car or truck ever built, and when you can do this, you can fix anything.

What kinds of tools? Well, you're going to need a complete set of box-end wrenches; all kinds of pliers and screwdrivers; several saws (including a power-driven saber saw, if you're going to be anywhere near some electricity); drills; a good soldering iron (and some knowledge about how to use it); three or four hammers; chisels for both wood and metal; and lots more. If you're totally ignorant about such things, go back to your library or the nearest magazine stand and find some books on home, TV, and auto repair. If you're lucky, the library will have back years of such magazines as *Popular Mechanics* and *Mechanics Illustrated*. These books and magazines are written for amateurs, and they begin at the very beginning. But you might as well make up your mind to learn some practical things if you expect to drop out successfully. And if you're going to live off junk, you have to know how to fix it. Getting good tools is step one. You can also go down to your friendly hardware store or Sears Roebuck hardware department and look around. Most (but by no means all) clerks know something about their stock—look, ask, and listen.

Subsystem Mysteries

Old cars, like any other piece of mechanical or electrical junk, are junk because some subsystem or another isn't working right. The whole mess is quite logical, in spite of all the weird things that seem to be under the hood, behind the dash, and everyplace else. What usually happens is that one subsystem is really fouled up, like maybe the carburetor is shot. So the thing runs pretty badly, if at all. If it doesn't run at all, incidentally, you can probably get the car for nothing for hauling it away, or maybe five to ten dollars. The other subsystems are sort of worn, but they usually work.

Now, your problem is like a doctor's—and cars are a lot like people, except that you don't need a license to work on a car. There is a breathing system (the carburetor and related parts); something that makes the car move (the drive train); a speaking/receiving system (horns and radio); and so on. When you look at those auto-repair books, you will find discussions of such things as electrical systems, fuel systems, brake systems, and so on. Each one is tricky, but logical—and anyone who is strong-willed enough to drop out can master this sort of thing. If you think that all of this is a waste of time, welcome to some distant marginal farm, where you have to work with your human muscles all the time, and where no one has any time at all to do anything—survival is more important, and very difficult. Perhaps the sad part of modern life is that the whole world is so wrapped up in mechanical gadgets that it is virtually impossible to live better than a fourteenth-century peasant without them. So, you might as well sigh, roll up your sleeves, and get to work. Take your time—if you drop out, you will have more time than anything else.

Now, your first step is to get your junker car working right. If you've spent a few months finding out how to buy the right one, you may be surprised at how easy it is. Cars are not like the one-hoss shay—they don't totally fall apart at once. The reason your so-called junker is wheezing and gasping is that some minor thing is wrong, like it hasn't had a set of new plugs and points for five years. You can learn to change these in any car in a day—and suddenly the thing isn't a junker anymore. But old stuff does break down, so keep learning. If your brakes work fine now (and learn how to check them out), they will need work next month or next year. Maybe you'll need a whole engine—but that's why you should talk to junkmen. With luck and some looking, you don't pay $500 plus for a new engine—you plug in a perfectly good old one taken out of a wreck, and for a hundred dollars or so you're good for another five to ten years.

You can start twinning your car, as we mentioned in the last chapter, if you please—it's an easy and cheap way to keep a car running. But don't do this until you know that you're going to stay in one place a while, since it is very tough to move a half-stripped twin even a block, let alone a few hundred miles.

This junk business, if properly analyzed and evaluated, and if you think ahead, can lead to really big savings. If you blow a tire out in the middle of Nevada, it can cost you $100 for a replacement at the remote gas station which may have one. But if you take a close look at your tires before you start, see trouble coming, and plan ahead, you can go down to a local junkyard and buy three-quarters-new heavy-duty tires for two to ten dollars each. This is the sort of thing you will have to know a lot about if you are going to drop out, so go to it!

Safety—First, Last, and Always

A safety note is quite relevant here. Bad brakes can kill, as can a tire that looks good but isn't. Even professional mechanics make mistakes, and fixing up old cars without carefully checking out all your work is *very* dangerous. So, until you *really* know what you're doing, have someone who does know check you out. Some of the most dangerous cars on the road are ten-year-old junkers that haven't been fixed up in a long time. But some of the safest cars on the road are those same ten-year-old jobs that have been carefully maintained where it counts—they do have good brakes, wiring, lights, tires, wipers, steering systems, and horns. The police sometimes assume that anything old is dangerous, which isn't quite true.

I have had my brakes fail completely five blocks away from the professional garage that fixed them—and I have run 20,000 miles on brakes I fixed myself. It's not who does it, but how well it's done that counts, and one of the nicest parts of this fixing up, once you get to know what is really going on, is that you are quite sure nothing will fail, because you did it yourself. Right?

I have also noticed that when you do the job yourself a few times, you know pretty well what you might have done wrong, or what might go wrong. After someone else has worked on your car, you can take it out on the freeway and cruise at 75 mph all day, not knowing that a cotter pin or bolt that might not have been stuck in right is the only thing between you and disaster. But if *you* stuck that cotter pin in, you tend to think about it a bit, and take it easy until you are sure. It doesn't take too long in working on old cars to realize that lots of things can go wrong until you do everything right, and this knowledge alone can be a sobering thought. So be safe, if nothing else.

Also don't go putting things on your old car that weren't there to begin with, without checking your state law first. Remember, we are in a safety-ecology, buy-new-all-the-time period, and the state laws that are supposed to protect us often make it very difficult to fix things up right. Some states have laws that prohibit using tires wider than were on the car when new, which means that the modern, safer radials often cannot be used. The law is intended, one suspects, to protect us from mad hot-rodders, but the effect is to make other recyclers less safe. Of course the companies that make and sell new cars are delighted—with a bit of luck, they may someday get laws passed that ban all old stuff, and we'll have to buy new every three years. Great for them, but lousy for ecology, conservation, and energy use.

But here the much abused, often debated Fourteenth Amendment comes in handy. Remember that no one can be deprived of property without due process of law, so if you keep your 1914 Reo in original mint condition, you can still drive it on public roads, no matter what safety standards have come in since it was built. Laws typically have grandfather clauses in them, which means that anything that was once done legally is still legal, even if it isn't for new machines. Stay stock, and you are safe.

Here we run into the usual American problem that anything old is definitionally bad, and should be thrown away by tomorrow morning, if not sooner. This is just one more example of how our society works, not only against junk, but also against lower-income groups who can't afford to buy new all the time. But what would you expect in a world run by upper-income, new-oriented people?

Indeed, a good way to start learning how to be a long-term, happy dropout is to rework your whole junker from front to rear. It will take time, but in the end the payoff is there, in part because if you can do this, you can do lots of other things with non-auto junk that you will need to do.

Incidentally, if you figure to hell with it at this stage and just stay in the System, you will discover that your income has gone up a thousand or so a year with the auto knowledge alone, or have you looked at your last auto-repair bill? Most people with money buy a new car every few years, in which case your $500 to $1,500 depreciation cost per year goes right on. But if you fix your own, you don't have any depreciation, since any car that runs is worth a hundred bucks, and that's what you have. Maybe the satisfaction of beating the game even in this modest way may be enough for you to stay in the rat race. You can console yourself with the thought that by doing this simple (?) thing, you have beaten the motor interests, done your modest bit for ecology by recycling a car that would have been junked, and learned how to do something satisfying for yourself that is a bit out of the rat race.

Location Strategy

Now you have your car, step two is to figure out where you want to be. It seems a bit silly to go where everyone else in the rat race is, like Los Angeles, but it's your choice. Most dropouts prefer warmer weather, so anyplace where it doesn't snow is fine. It saves lots of money on heating bills, and few people like to chop wood for fuel, or shiver in unheated shacks. But wherever you go, remember that the best junk is near or in nice split-level suburban developments inhabited by high-income UMC types. So plan to be within driving range (say twenty-five miles) of such a suburb. There are lots of nice places of this sort—near Florida cities, around (but not in) Atlanta, near many Texas suburbs, and across Arizona, New Mexico, and California. It's worth noting that some of the best drop-out country around is in the lower Midwest, simply because no one ever thinks of dropping out there. It gets cold, but the area is full of decaying, slowly dying agriculture towns, with lots of very good cheap housing. And the big, juicy suburbs surrounding the major cities are not far away.

Meet the People

Now, the next step seems peculiar, but it's essential. You have to learn to talk to people, on their terms. This is hard to do for most drop-out types, since they have been subtly brainwashed all the way through their good universities to have total contempt for the yahoos who get work done; but remember, these fellows (it's a man's world out there, so few women are involved) are the ones who control junk. They either own it, pick it up, or, very importantly, know where it is and when. A short friendly chat with a guy who runs a washing-machine repair shop in some dreary town no one ever heard of can turn up more usable junk than you would believe—if you talk to him about what he wants to talk about in the right way. He could care less about the burning issues of the day (though you will find some interesting surprises from time to time on this point), but he sure will know where the junk is. He may well have a few tons of it himself out back. As you start dropping out, you have to learn to talk to everybody—the milkman, farmers, the fellow who fixes power lines, the store clerks, and all the rest. Remember, you're going to be living off the land, and these people know the land and everything that's on it, in it, or below it. And you may discover, to your surprise, that lots of these people are fun.

Housing the Successful Dropout

So, you go to your promised land. Now, how about housing? If you're like lots of young affluent dropouts, your vehicle is a VW or Chevy van, and you live in it for a while. Those things can carry lots of stuff, plus some sleeping bags. The gas stations and rest stops on the freeway have the water and toilets, so just live there. But if you are finally deciding to put down some roots, get to the town or city, and drive around. Drive around a lot. What you want is not some real estate agent's idea of a dream house; what you want is a solid, but decaying wreck that has possibilities. In a country that throws away everything else, why not houses? There are over 500,000 abandoned American houses, with perhaps another million or so that are all but abandoned. And something over 2 million residential house trailers, or mobile homes, have disappeared from the state registration lists in the past decade.

Where are junk houses? Like other kinds of American junk, these are things that are pretty good, but have some fundamental defect. They tend to be where people don't want to be, but if you're dropping out, this should be no problem. Most of them are in inner-city areas, and they have problems like leaky roofs, shot plumbing, cracked plaster, and wiring defects. Inner-city houses tend to be in neighborhoods where even strong men hesitate to walk at night, but once in a while you can get one that is reasonably located.

The best junk houses around are those in smaller cities and particularly towns where everyone else has moved away. Old coal mining towns have many, as do Midwestern and prairie towns that used to be agricultural centers. As everyone else is moving into the city and suburbs, causing you to think about dropping out, they leave behind lots of nice old houses in places where no one can make a decent mainstream living.

Such places are great if you want to write or paint—they tend to be nice quiet towns and neighborhoods, with lots of older and quieter people around to keep you company. You can get a lot of work done in such a place if you don't really want to be where the action is. But that is exactly why you are dropping out. And these old houses have lots of room around them for twin cars and piles of good junk.

There are still plenty of abandoned farmhouses, too, and in some neighborhoods, in the Midwest for instance, you can even find them along the Rural Electrification Agency power lines. One near where I live is already wired and has an electric water pump and an indoor toilet. You tend to find that as the number of farmers decline steadily, they abandon houses but not the land. If you live in such a place, be prepared for some eager neighbor occasionally plowing around your doorstep on the good farmland that is still in use.

House trailers and mobile homes are rather like cars—they wear out fairly fast, and they begin to look pretty scruffy after a few years of neglect. The outer panels get dented, and the upholstery and floors inside get beat up fast. But you can buy a ten-year-old mobile home for as little as a thousand dollars, or even less, if it is really battered, and you can move it anywhere you want. Remember, though, that you will need a pad with plumbing and an electrical hookup, unless you really are roughing it.

Now, if you can fix up a car, you can fix up a house—it's harder physically, but easier in skill terms. So get in there and paint, sand, rewire, do the carpentry, and all the rest. This sort of thing is getting to be so common that it hardly qualifies as dropping out, but it is lots of fun. Talk to those who have had the satisfaction of converting a real wreck into something desirable. Older houses in particular offer real esthetic pleasure. You laboriously peel off ten or fifteen coats of paint, and suddenly all that lovely old wood looks like it did at the beginning. But even rebuilding an old mobile home can be an adventure in scrounging and craftsmanship.

Now, if you can master fixing cars, you can master the crafts necessary to furnish your house. You need to know carpentry and plumbing of course, and have the relatively few good tools to do the job. But for the consumer-goods package of washing machines, heaters, driers, vacuum cleaners, power drills, refrigerators, and all the rest, you really need only two additional basic skills. One is how to do fairly simple electrical repairs, since most things get tossed away because they fail electrically. The other is how to work with refrigeration and heat-exchange equipment, which involves freon-gas work and some basic physics skills. You can go a long way for very little knowledge here, since in many cases parts changing, rather than rebuilding or repair, is the key. The washing-machine motor is burned out—instead of rewinding it, just find another motor and bolt it in. And all those tools you bought for your car, plus a few more, will do anything you want to do.

Remember your carpentry skills for the furniture, and add some simple principles of bolting, riveting, and gluing things back together. Steel furniture gets tossed out because it breaks at a weld—a bit of scrap iron, a 1/4-inch power drill, and a few bolts will put you back in business. Wood chairs lose legs—figure out how to glue or screw them back together. It is really surprising how far a few very simple skills can get you if you have time to work on things. And isn't that what dropping out is all about?

Finding Your Junk

Where do you get the stuff? Mainly from that posh UMC suburb on trash days, when the rich folks toss away everything. Just find out when it happens (by talking to the trash people and garbagemen), and get there ahead of the trucks. It's all free. You can amuse yourself by learning enough basic electronics (which, all talk and warnings to the contrary, can be learned in a few weeks by duffers, if they are willing to study a bit at the public library) to fix up the abandoned TV sets and record players as well. Or, once you get the things working, sell them to buy some of the finer things in life. And, if you don't mind waiting a bit for your reading material, the same trip can keep you up-to-date on the world, a few months after it actually happened. Remember that most magazines and newspapers get thrown out, too.

Sometimes you need something now, and the junk pickings are thin. There are other places to go to find it, although it might cost you some money. Auctions are very common in many areas. A farmer will retire and move South. Before he goes, he will sell everything at a public auction. Drive out and see what's there—if it's a nice old farm, you can find anything from antiques to pure junk. It all gets sold to the highest bidders, and since most people there are either furniture dealers, antique hounds, or farmers looking for equipment, you can often find real bargains. If you're lucky, it will be a cold or rainy day, and few bidders will be around.

There was a huge estate auction of just such a farm last summer near here, and it was fun to see how many people have already picked up the word on junk. We had three kinds of beards down for the auction—the neat square cut ones of the Amish, who came down to buy the horse-drawn farm equipment offered (it was an old and large farm, and the owner had never thrown anything away). Then there were the well-trimmed vandykes of the college professors, nosing around for antiques (and there were lots around). And finally we had the scraggly beards of the local commune, looking for bargains in old pillowcases, comforters, and blankets (which they got). Anyone willing to buy what others didn't want, like used and banged-up but serviceable kitchen pots and pans, bedsteads, various assorted kitchen chairs, and slightly ripped sofas, could have equipped a five-room house complete for under $100.

The stuff you could have picked up was so weirdly out of style as to be in style. Who else would have a dinette suite with every chair different, or a very nice easy chair from 1934 that didn't match the sofa from 1948? But if your only interest is comfort and beating the game, a good auction is hard to beat. If they happen in your area, drop in. Incidentally, at auctions you can sometimes get very good tools cheap, particularly the odd ones that aren't used too often. Does anyone want, cheap, a semiautomatic valve grinder for a 1925 car?

Another place to go for you city people is the local St. Vincent de Paul or Goodwill stores, or any of the other charitable organization operations. Here many people pass on their usable stuff, particularly clothes. The Davis California Goodwill shop sold me the best sports jacket I ever owned for fifty cents—it cost over $150 new, and it had only been worn once. Prices are usually very good, and big city stores particularly have surprising varieties of stock. Books are a drug on the market, so if you like reading, go and browse.

If you still haven't got the hang of fixing appliances, try these stores for low-cost renovated merchandise. Goodwill, as one example, trains handicapped workers in appliance repair, then sells the output. This material is not quite junk, but a cheap way of beating the game, even if you are unwilling to do all this learning I've been talking about.

You also might seriously investigate how Goodwill and similar organizations train their workers. If you're going to live off the land, find out how an efficient organization does their training in all the things you'll need to know. And, after you see what kind of fantastic job these dedicated people do with some very badly handicapped people, maybe you'll decide not to drop out at all.

A good place not to go, but one where everyone ends up, is a used furniture store. These outfits usually cater to very low-income people, often on real easy credit—like 500 percent per year interest on the balance. The stores always have what you want, though—that's how they manage to survive, and their owners have plenty of years of experience in finding out just what most people want. And once in a great while you can find a bargain. But this is really not the junk business.

Read carefully the really dreary parts of the want ads in the paper you can read at the library or pick up in the bus station. Often a government unit will have auctions on things. City police departments sell abandoned cars (though in many states such cars can't be titled but must be scrapped, so check first). Schools sell old furniture, as do courthouses and other units. You can get anything from typewriters to bulldozers cheap this way. By law in most places, such auctions have to be advertised well in advance, and the goods can be inspected. My brother has been buying California Highway Patrol cars this way for years—he claims that the cars, even though they have a half-million miles on them, are so beefed up and well maintained that they're a bargain. Maybe he's right—at least it keeps his transport costs down. I happen to be writing this in Toronto, and I notice that next weekend my borough is going to auction off a couple of 1970 Mercurys. I guess I can go down and take a look, anyhow.

Everyone knows about war surplus, and the U.S. government does sell off more odd stuff than you would believe. But this is a national game played by experts, and unless you just happen to be in the neighborhood when a big auction takes place, your best bet is to go down to the surplus store and buy it retail. It is less time and trouble. Most military hardware, including trucks and jeeps, is so overbuilt and expensive to fix after you buy it that it is usually easier and cheaper to find something civilian more or less like it. Gold plating is fun for the military, but it's tough to fix for us poor civilians.

Another Midwestern custom (now becoming common throughout the U.S.) that leads to good cheap stuff is the garage sale. One or more families will pick up their good junk, advertise it in the local paper for sale over a weekend, and wait for business. In our city of 50,000 people, there are an average of ten per week during good weather—and some have been going on continuously for years. Like used clothing and furniture stores, some are good and some are bad—take a look before you leap.

City dumps used to be great places for finding odd stuff, and maybe some of them still are. But in many localities, ecological concern and fast dirt filling over rubble has eliminated the good trash as well as the rats. Out in rural areas, in spite of "No Dumping" signs and draconian penalties, you can find informal dumps just about everywhere, and quite a few of them have interesting stuff like sofas and old coal-burning stoves in them. Also in rural areas you can find more abandoned cars than you can handle, unless you want to get into *that* business. You can also occasionally find real goodies like a dump of used lumber (common around major construction sites as well—lots of times you can get all the plywood and two by fours you need to fix up that house for nothing, if you bother to nose around and talk to the right people. And what doesn't go into your house or trailer will make good firewood).

Some Junk Isn't Junk

When you've gone this far, you will have discovered that junk changes through the years. A lithographed Coca-Cola tray might have been made in 1921 for a few cents and given to a soda fountain as an advertising gimmick. By 1930, the thing was tossed casually in the junk pile. By 1960, some people were beginning to pick up such artifacts as interesting Americana. By 1970, the trays were antiques, worth anything from a few dollars up to a hundred or so, depending on how rare and/or desirable it was. Some of today's junk will be tomorrow's antiques, and the problem is to figure out which junk will be worth something in the end. The usual furniture, cabinets, and arms (guns, knives, swords, etc.) are supplemented by various forms of barbed wire, canning jars, very old radios (to 1930), telephone-pole glass insulators, farm implements, toys, and lots of other artifacts, which often are not very old. If you do find a box of 1914 Ball canning jars, you might want to use them to put fruit in, but now they may be too valuable for everyday use.

You can check this point out easily enough by going to various antique stores and talking to owners. They know very well what is worth something and what isn't, and if you are going to spend a lot of time messing around junk, particularly junk coming out of older areas, it may pay to know what to be looking for. Good newsstands carry various specialty magazines about antiques and what is wanted, so buy a few and get educated. The odds on your finding a 1923 Atwater Kent radio in good condition on top of a modern junk pile are low, but who knows? If you do, it may be worth hundreds of dollars. And don't make the mistake of thinking that anything old is valuable, because it just isn't so. Someone has to want it, too.

You Can Eat Some Junk, Too

Now, you have to eat, too, and even this can be managed in many places. Do some research and figure out the food codings, particularly for frozen foods. All those mysterious numbers mean something, most importantly for you, the throw-out time. Under sanitary and health laws in many states and cities, stores must discard such items after a stated period. The code tells them when this is. Now, the law says the stuff must be dumped and destroyed, but if you talk to the right man, and if you get to know the right times, you can have your own clean garbage cans at the supermarket back door when the stuff gets thrown out. Remember, the safety aspects of the food coding mean that the food is safe at the expiration date—it has to be. Something that will last perfectly well for 120 days has to be dumped in 90 just to be on the safe side. Take the stuff back to your own freezer, toss it in, and enjoy yourself for the next month for free—less the electric current needed to run your freezer. Where did you get your freezer? Well, freezers that don't work are a real drug on the market, and you can normally have them for nothing. Even the old ones that work are cheap—I've seen them go at auctions for under ten dollars. And if you just happen to know a little bit about refrigeration equipment, you can fix up a couple and have all the refrigeration you want, even while dropped out.

The frozen-food racket leads to some odd gourmet habits. A friend of mine out in California who does it found himself living on imported frogs' legs for two months—you take what no one else wants using this system, and that tends to be the odd items.

Now, if you really want to live in style, find the day-old

bread place, and the one store in town that sells rice and flour in hundred-pound sacks, for about one-fifth what a five-pound package costs. As for booze, I think I know where a couple of bootleggers are around here, but that's illegal of course. Try picking up some empty whiskey bottles with the very best labels (there are *lots* of these in trash coming from the uptight, organization-man split levels) and filling them with the cheapest booze you can buy. A lot of people can't tell the difference. I suppose this is illegal too, but I doubt that there has ever been a conviction for it. Then you can entertain in style.

Portrait of the Successful Junkie

Now you are ready to drop out the way it should be done. Cheaply, comfortably, with the feel of beating the game totally by just doing what everyone else isn't doing. You sit in your oddly matched furniture, watching your 1958 Zenith TV you fixed up, sipping your rebottled booze, in your junk house you painted and repaired yourself. Your 1961 Ford station wagon is outside, ready to take you where you want to go. And the whole thing cost you a few hundred dollars, plus just lots of time.

You have lots of new friends, kooks and solid citizens, because anyone who can fix anything is going to be very popular in this world. And you can fix it all. You have the time to read, to think, to do simple things like fix a chair or paint a room, or write some poetry (get the paper from any computer center—the back side of computer printout is unused, and the stuff is generated by the ton). You see, the world is full of mechanical and electrical things that are all just simple logic to get right, yet the way we've got the world structured, no one is encouraged to fix things. Toss it out, buy new—why? It's more fun to make do.

Now, all this work and trouble may seem too much for

one who just dislikes the rat race. But dropping out really involves two possibilities. The first is to do nothing useful and become a parasite on the rest of us. There is nothing particularly wrong with this, except that it is dangerous. We already have enough half-baked philosphers, thinkers, and poets around, so there is no particular incentive to support more. Parasites have a bad habit of getting sprayed, or something.

The second option follows along the lines suggested here. Instead of being a parasite, you become a useful citizen, living off others' droppings. Not only do you aid our ecological movement, but you acquire useful skills. Moreover, you end up not dependent on anyone else—indeed, if you really learn to fix things, it is likely that others will be dependent on you. And since no one in authority pays any attention at all to junk, you're *really* free. After you learn the ropes, you will find that you can live a happy life working only a few hours a week, leaving the rest of the time free for whatever interests you. And since there is plenty of junk everywhere, you can live just about anywhere you please.

You still have to buy electric power, water (maybe—well-digging is still not a lost art), and other utilities, but so far these things are cheap enough. And you have to worry about medical services and similar personal things. Try welfare, which is quite a different story from this book. One curiosity about welfare, however, is that the system is designed and administered by those in the mainstream, the same UMC types that inhabit the split levels in the suburbs you're scrounging in. They build their own fantasies into the system. Hence, you can live awfully well within the system if you know how to scrounge junk, since the system has nothing like that built into it. It assumes (maybe correctly) that most people need the consumer-goods package, including nice houses, TV sets, and all the rest (except maybe cars), so they have to have that kind of income. A smart scrounger can take his welfare checks and live the way he wants.

Junking the System / 79

If you bother to sit down and figure out the market value of all the things you can easily scrounge in this country, you will discover, to your surprise, that your *real* income is likely to be well over $10,000 a year. This follows from all the work you do for yourself. You find a friendly junkman to get a carburetor from, install it yourself, and you're out maybe $5. Your UMC equivalent will take his car to the agency, get a new one installed, and pay a bill of perhaps $100. In effect, you've just made yourself $95—and when you add in all the gains from fixing freezers, toasters, and all the rest (and maybe some real gains from selling some of your items), the sums can add up.

Moreover, and very importantly, your income is tax-free. A typical UMC type in the 40-percent tax bracket has to make about $150 to pay his $100 bill—your $5 job was tax-free. As taxes get higher on income, which includes not only the income tax itself, but also social security and local payroll taxes, the possibilities of barter become very interesting. If you can *really* fix all those things, try getting down to the UMC suburbs and bartering some medical care or legal talent for a car repair or getting the vacuum cleaner to run. You both gain tax-free income, and everyone is better off.

Junk, Welfare, and Ethics

This junk-scrounging business is one reason why so many middle Americans are so uptight about welfare. They know full well that any reasonably intelligent operator can double his income by doing things himself and scrounging things off the economy—many of them do some of it, and they all know people who do lots of it. They take a look at the size of the welfare checks, designed to keep barely alive a welfare mother in some cruddy ghetto, and figure out what they could do with that money out where they are. Then they get upset and start mumbling about the welfare chiselers. But, given the total lack of interest in junk by the UMC establishment types, these middle Americans will never be understood. If Mr. McGovern had understood what this chapter and the last one were about, he might have made a much better showing in the 1972 election.

One recent study suggested that the typical American family did about $4,000 worth of work for itself every year. This is the market value of doing things that could have been paid for, like doing the laundry, driving the kids to school, preparing food, and sewing up rips in the kids' pants. If you drop out and get to be a good junkie and repairman, that $4,000 might be the $10,000 we suggested, or a lot more. And remember, even if you're honest, it's all tax-free. You can earn a hundred bucks at some dreary job, pay twenty of it in taxes, and buy a cheap sofa—or you can just find a good sofa at some dump or UMC house for nothing, spend ten hours getting it in shape, and come out ahead, with no taxes to pay. After all, you didn't earn any money, did you?

Everything said here rubs against the present mainstream, and in talking to UMC people about doing some or all of it, I get the feeling that this is all very non-U. People are not supposed to do such things! We still keep telling our smartest young people in our best colleges and universities that to work with your hands is degrading and nasty. Curiously, the most radical types tend to be the ones who have the most contempt for the working peasants—more business school types fix their own cars and are proud of it than the theoretical Marxists over in the arts college. But thinking is what it's all about and where it's at. Then, when thousands or even millions of kids seem to want to drop out, we wonder why. We've killed in them half the fun of being in a technologically oriented civilization—the half that involves figuring out how things really work, and what kinds of engineering and craftsmanship, if any, went into the gadgets. We pseudo-replicate this urge to know with craft shops and basement workshops in the split levels, but why not get out and do it the real way?

Moreover, this sort of thing is coming on fast. We're running out of energy, out of materials, out of everything, and it seems clear that within the next thirty or forty years, lots of things are going to get recycled. Why not begin now in your own style?

Yet, when I talk this way, I sometimes think people may be listening. A surprisingly large part of *The Whole Earth Catalog* deals with maintenance manuals, how to fix things, and similar stuff. The very few successful communes around work just about this way, with members learning craft skills to keep things going. And in some circles, having a ten-year-old car is kind of chic these days, and if you work on it yourself, you're not an outcast, but a man to talk to for advice. People all over are grabbing up the lovely old houses in the inner city and laboriously restoring them, and they aren't exactly dropouts either. Liberal intellectual types are talking about recycling, and here and there I even detect a slow return of the healthy respect we used to have for a fellow or girl who could do skilled handwork.

But it's still not chic to get an old TV set and fix it the hard way, with a manual in one hand and a soldering iron in the other. It's not too chic to get yourself covered with grease, dirt, or some other even gooier crud that tends to be around trash. And a new car, new house, new everything are still the mainstream success symbols.

But ignore the mainstream. You can drop out the way we suggest here without going anyplace, giving up your job, or whatever. Take any piece of the thing—find yourself a beat-up old table, or a toaster that doesn't work, or a recalcitrant car. Do some work instead of agonizing over the plight of man—get to the library and find out how things tick. Then fix it yourself. If you can't stand machinery, learn to sew and try retreading old clothes or blankets. And someplace in the middle of this trash and junk business, you may find, to your surprise, that you've found a piece of yourself that is very valuable and that you never thought you had.

The trends are mixed. This sort of thing has been going on all along, out there in Archie Bunker land (it's interesting that Archie can't fix anything—most guys in his position can do lots of things, but then the typical script writer, even a brilliant one, is not likely to know about such matters). But at the same time, all our norms and values have pushed the other way, to the point where nothing seems to make much sense anymore. Maybe, in the end, we are all uptight Calvinist string savers, people with some faint instinct for craftsmanship, and we really want to live off the land like we thought we used to. At least it's worth trying.

4

Making Money on Junk: Autos

About six or seven miles north of Bloomington, Indiana, on Highway 37, drivers can see a sight repeated ten thousand times across this land—the classical auto junkyard. Stretching across a couple of hills are thousands of wrecked and junked cars, in every possible condition. Some are virtually stripped hulks; others are almost complete. There are a few buildings and sheds here and there, a hand-painted sign announcing the name of the present owner, and greasy piles of this and that everywhere. An ancient wrecker truck, probably scrounged from the bones of half a dozen others, growls around the yard, and a couple of men torch off a drive train here, or pile up some engine blocks over there.

The neighbors and citizens are always up in arms about this place, because, even though there is a good fence around the property, it is very visible from the high road. They are always trying to get the county to get rid of the yard, to get those reminders of nice cars out of the way someplace, somehow.

And if you go in to the junkyard, you will find a friendly enough operation. In the old days, a few owners back, you could ask for your part, say a generator for a 1961 English Ford, and the owner would think a minute, scratch his head, and say, "I think there's a Ford over there, down the hill, go see if it still has a generator on it."

"How much do you want for it?"

All the while you've been talking, the owner has been sizing you up. He takes a look at your greasy pants and oily hands, which are OK. But that foreign accent has him bothered. "Two dollars?"

The Ford agency priced you one new for $29.95, but you're getting smart in your old age. You hesitate, rub your chin. "Has it been covered?" Meaning, has the generator been out in the rain, or under a hood or something.

"Don't know. Tell you what, how about a buck and a half, and you put it in. If it doesn't work, no deal."

"Fair enough," you agree.

"Got some wrenches?" he asks.

"Yeah, out in the car."

"OK, go get it."

You wander down past rows and rows of other people's dreams. Who would have ever bought *that* monster new? The old luxury car is battered and torn now, with the whole front end missing. And close by is a real wreck. You can glance at the shattered, twisted compartment and wonder what the driver hit. Whatever it was, he didn't come out alive. Close to that one is a fairly new foreign car, whose sales have been dropping like a rock, after the first few big years. Too bad, the car just couldn't stand up to American freeway driving conditions. . .

The English Ford is there, with the hood loose, but still hanging over the engine. And the generator is there too, so you take the wrench and remove it easily—there are only two bolts, plus the electric-wire connection. And you take it back and replace it in your car, start up, and it works. You pay the owner a dollar and a half—he's supposed to collect sales tax, but he doesn't have the change, so he forgets it, and you're back on the road in good shape. And, as you leave, you see the next customer start out to the field, looking for a front brake drum for his '66 International pickup.

Everybody sues everybody these days, and not every person is really honest, so now when you go, you pay more, and the owner doesn't let you out in the yard. The workers will get the piece for you. But it's still pretty much the same old operation, with lots of old junkers sitting around being slowly stripped for parts.

This is the image most people have of modern American junkyards, but actually, these things are only one small piece of the total picture, as we shall see. This is just the tip of a large, profitable iceberg.

In the totally affluent society, it sometimes seems that we are buried in junk. Autos, being big and ugly when they are beat up, are the evidence—we find them everywhere, and somehow we never seem to get rid of them. Over six million cars are scrapped each year in the United States, and even a short ride in any city or countryside would convince the casual observer that most of them are piled up right where he's at.

Every city has an abandoned-car problem. The things are worth so little that you have to pay to haul them away, and it costs the city money to do just this. When enough of them are piled up, they are usually sold at auction to junk dealers. But not every car is abandoned in the city. Country lanes have their full share of rusting hulks, too, except that in lots of places they never do get picked up, and they sit wherever they are, sometimes for years.

Yet, there can be money in junk, as many junk dealers can tell (but won't—why mess up a good thing?). To casual observers, a car is a car is a car, but to a junk auto dealer, every hulk is something special. Some are worth virtual fortunes, while others you have to pay to get the junker to take. And because a car is not a car, we have problems, along with money-making opportunities which anyone can take advantage of, if he wants to take the time and trouble. And that is what this chapter is about.

Scrap auto markets are among the most highly organized in the world. If you want to see how a really pure capitalistic system works, you could study the industry with profit, for it has no status, and hence very little regulation. About all that need worry you in the business are zoning codes and antipollution ordinances, since most people don't like auto wrecking yards next door. But you can buy what you want when you want for whatever it takes to pay for it, and sell the pieces the same way. And, like other complicated businesses, there are both idiots and geniuses in the game. Most junkies barely scrape by; others drive almost new Cadillacs. It all depends on how good you are.

Auto Recyclers—The Pros

There are actually many businesses in one in wrecking cars. One that is desirable is the relatively new car business, How does it work? Well, suppose your cousin rolls his new Buick and totals it. The insurance company takes a look at the wreck, sighs, and pays off. When it pays off, the company owns the car. It sells the car to the highest bidder, which will be a wrecker specializing in this late-model stuff. The price will depend on the shape the car is in, plus the probable miles it had on it when it rolled. Now, if the Buick was last year's model and had only 8058 miles on it, it is likely to go for a very good price, perhaps as high as 80 percent of what it was worth new.

The reason is simple—there is lots of gold in that twisted hulk. The wrecker will quickly disassemble the entire vehicle and begin to sell the parts. That hot V-8 your cousin was so proud of may go to a hot-rodder, or maybe to a garage that has another General Motors car with a shot engine. With new engines going for over $500 these days, and with one lasting over 80,000 miles, that 8000-mile engine can be worth $400 or so. In the last chapter, we noted that you can buy a good engine for $100. You can, for an eight-year-old car, but not for last year's model. Read on to see why. The whole drive train (transmission, rear end, drive shaft, etc.) may go the same route, since cars under three years old get fixed, and often it's cheaper to replace a component with an almost new one from a wreck than to rebuild the broken one or buy new. And with so few miles on the wreck, it's a good bet that the parts will be in good shape.

For those of you who can remember the days when they built really good cars, back in 1925, you may recall that you reground the valves at 5000 miles; rebuilt the engine at 10,000 miles, and probably replaced the whole drive train at under 20,000 miles. But these days, when most (but not quite all) engines and drive trains are expected to go 80,000 to 150,000 miles without wearing out, the almost new cars are worth lots, because they still have much life in their parts. Cars aren't like they used to be—they're a lot better, and wreckers know it.

Now, if your cousin happened not to bend up the front end, that piece is also worth a tidy sum. A little old lady in the neighborhood wrapped her new Buick around a light post and scrunched up the front end something awful (or maybe it was a drunk at 2:30 A.M., which is the real danger hour for auto accidents). The insurance company that has to pay will get some estimates for repairs. One way to fix it is to have a highly skilled (and expensive) body man carefully pound out all the dents, but another way is to just buy a whole new front end and bolt it on. When painted, the car will look like new. So the low bidder on the job contacts the junkyard, buys the front end, and within a week the damaged car is back on the road.

The same thing will happen to everything else in the car. Windshield wiper motors, all the glass that isn't broken, the doors that aren't bent, the instruments, and so on will be taken off, cataloged, and sold. Things liable to collision damage, like front ends, trunk lids, doors, glass, bumpers, fenders, and similar parts will be most valuable, along with expensive motor and mechanical parts. But from time to time, some gismo or other will fail in a relatively new car, and if you have the right almost new parts, you're home free.

In the end, what's left is a hulk of bent steel, and it gets tossed in the furnace for scrap. The dealers in this business have teletype systems, and lots of information about who needs what. Some kinds of cars have big damage for little bumps on a front fender; if so, undamaged front fenders on this model will be high priced, because there will be a demand for them. The body and fender shops keep in touch, as do all the repair shops. Even the General Motors dealers will be in the game from time to time for odd parts, since it may be easier to install a used part you can get today rather than wait a week for a spare. When my three-year-old Olds burned out a windshield wiper motor, the friendly local junkyard just happened to have one off a wreck, and it's in there now, doing its job. The dealer didn't have the new part, and rather than get me mad for a week, it got me mad for a day by putting in the old one. Our local dealer is ethical—he told me about it and adjusted the bill accordingly (downward). But not all of them are so nice about it, and most customers couldn't tell the difference.

These junkyards are the kind that no one complains about much, largely because there is so little to see. You can find them down in industrial districts and in the yellow pages, often with the notation, "Late-model cars only." People are often mad and puzzled that these fellows won't buy their ten-year-old junker, or even pick it up if you give it to them. The reason is that hardly anyone fixes cars over three or four years old, at least commercially, so there is little money in having old parts around.

These operators usually take cars apart right away, so they don't have huge yards full of half-disassembled hulks. They even call themselves auto recyclers, not junkyards, which is what they are. When some local worthy screams about how the junkies are defacing the landscape, they take the heat just like everyone else. But normally, all you can see of their operations is a ten-foot fence around their property.

These fellows are also the most well-organized guys in the business. If you like growth industry, go try to get a job with one of them and learn the ropes. It's a tricky, complicated, and interesting way to make a living, but it's not for the ignorant.

Auto Junkyards—The Hard Way

Cars over four years old and on down to maybe fifteen to twenty years old form a different market, and are handled by a different group for the most part. These cars do pile up in huge yards, usually fairly far from the city, since they need lots of land to spread out in. This is one reason why you can't give away an old car in a city—it costs too much to come and get it. The boys will buy it only after someone (like the local police) have assembled a bunch of the junkers in one place, where they are easy to get at.

This market is organized, too, but not quite as tightly as the newer junk market. For one thing, lots of smart people who own old cars come in and buy stuff themselves, so you get more of a retail business. Moreover, it's harder to figure, since no one knows when someone will decide to fix personally that bent front fender on his 1962 Chevy. But, if you have space, you can make some money, since you may have bought a '62 Chevy for five dollars, and if you can sell just a few parts off it, you're home free—if the rent isn't too high, and if you can take the local heat of the concerned environmentalists. Sometimes, if you're lucky, old stuff fits new stuff, as when Ford doesn't redesign an engine for maybe ten years. Such things as alternators, wheels, and various mechanical parts may be like this, so these things can be sold to patch up more modern cars.

But generally speaking, you can't sell it all, so much of it just sits there, with parts slowly being taken off for years. Finally, there's not much left, or what is left no one wants, and the whole thing is cut up for scrap.

Now a car is a funny thing. It has about $200 worth of scrap in it even if it doesn't run, but you can't give the things away. The reason is that the scrap is so mixed up. If the junker were pure iron or steel, it would be worth money, and we wouldn't have any sitting around. But it's steel, glass, copper, aluminum, plastics, fibers, and all sorts of things all mixed up in a hard-to-take-apart mess. And American labor is very,

very expensive. The result is that it hardly pays anymore to separate all the good junk from all the other crud. In a few places, the boys can burn off some of the junk, like the plastics and fibers, but because of smoke pollution this has been banned all over the place. The ban makes the wrecks pile up even more.

But in this free enterprise part of American life, if there's money in the game, someone will come up with a solution, and recently we have seen the development of huge portable (and fixed) car shredders, which work like your sink garbage-disposal unit. The whole car is shredded up, and various magnets, sifters, and sorters automatically sort out the various components. With copper selling for maybe 40 cents a pound, it's nice to get out the 20 to 50 pounds an ordinary car will have in it here and there. And doing this makes the steel worth a lot more as scrap, since copper contaminates the steel melt at the mill and leads to low-quality new steel. Similarly, the aluminum, brass, zinc, and whatnot gets sorted out and sold to be recycled. So far, this idea is just getting under way, but even small upward shifts in present scrap prices could lead to chewing up most of the older junk cars pretty fast. And getting scrap prices up might not be so hard, as we shall see. This is one problem that might be over before we figure out what to do about it.

Twenty-five Years to Glory

If a hulk manages to be hidden down in the cow pasture for over twenty years, a new game begins. Prices begin to go back up, since the collectors, who have the older cars, now are looking for parts, and the first two markets we noted above have pretty well stripped out the supply of wrecks. Indiana has it about right—any car or truck over twenty-five years old can get an antique auto plate, and twenty-five years is the magic number. An old wreck over that age rarely gets chewed up.

It gets back to phase one, where some specialist finds it, takes it apart, catalogs the pieces, and sells them (often on a nationwide market) to collectors looking for something.

So, if you find the top bows on an old 1936 Ford four-door convertible sedan, don't despair and wonder how you can get the junkman to haul them away—instead, put an ad in *Hemmings Motor News,* or *Cars and Parts,* which are two magazines that cater to the antique-auto market. If those top bows are in good shape, not too rusted or bent, they are worth a hundred bucks or more. And if you find something really nice, like a mint set of Duesenburg headlights, you may be able to pay off the family mortgage with the proceeds.

In auto junk, just as in non-auto junk, because something is old doesn't necessarily mean that it's valuable. Someone has to want it, and if the price is high, he has to want it bad. Moreover, the guy who wants it may be in San Francisco, while you just dug the junk out of the barn in Kansas, so transportation costs can knock a big hole in your profits. But it's a fun game, which you can see from time to time around the country at various antique-auto show swap meets. Here you find quite sober and serious citizens buying and selling the most ungodly pile of rusted old car parts you will ever see. And some of the prices will make you see why literally thousands of people have fun and make money too by finding really old cars in moldly barns someplace and get the pieces back into the market.

Like everything else in junk, you have to know what you're doing. But getting an education is easy—just start hanging around the nearest antique-auto club group, and within two years you will know what is worth something and what isn't. And anyone can play—so far, the government hasn't licensed such people, nor are there any price controls.

Your junk is my antique—it figures, in a society that has the time and money to fool around with all sorts of odd things. One warning: if you do start talking to these old-car buffs and finding out what kinds of junk isn't junk, you will probably end up owning a 1923 Model T Ford, or maybe a 1935 Packard sedan. And if it is restored, and runs, that kind of junk is worth thousands of dollars. Moreover, you can have fun just driving around in the thing watching the envious peasants lust after your possession. I know—I occasionally drive my somewhat beat-up 1931 Ford to work, and even though there is a Maseratti, a couple of Cad convertibles, and similar choice items in the parking lot, my only problem is to keep the curious off the thing—it's the one car there that gets any attention at all. If you like real status in the futurist American society, try going backward forty years or more. Then you can achieve *real* distinction.

The Rebuilders

Somewhere in the middle of this pile of junk is yet another market, which is that of rebuilding usable components. You can buy a rebuilt engine for almost any car made within the past ten years from places like Sears or any major auto-supply house. These rebuilders buy an old, worn-out engine from the junkers, then take it apart, replace or refinish all moving parts, and presto—you have a like-new unit, which is cheaper, yet will last as long as the new one did. Since it doesn't take much capital to rebuild, as compared to make-new, there are literally thousands of rebuilders of engines, transmissions, alternators, generators, starter motors, rear ends, and so on all over the country. If they do their job right, they make a real contribution to getting rid of the junk—you just keep on recycling the same parts over and over.

The difference between a brand-new car and a ten-year-old junker is about two pounds of metal. The junker has less metal—it's worn off the bearings and such all over the car, in hard-to-get-at places. If the key components are replaced with recycled, rebuilt ones, the car can be as good as new. This is exactly what an antique-auto restorer does—he carefully rebuilds everything back to like-new, so the car runs like new, even if it's forty or fifty years old. Anyone can play, but we Americans usually don't. We "need" the new cars, instead of being satisfied with a perfectly good, recycled 1963 Falcon, or something.

Making Money from Junk

I've talked to young people about careers for fifteen years, and so far no one has ever, no way, no how, ever raised the question of making a career from junk. This really is why we're in trouble with the stuff. All the prestige and much of the money goes to the producer, the fellow who makes something new. The guy who picks up the mess ten years later is a low-caste underdog slob. You can read the college guides to careers forever without finding one reference to garbage or junk, and no trade school I know of runs a course in how to be a better auto wrecker.

Because this is true, the possibilities are mind-boggling for anyone who wants to be in a growth market for the next fifty years. The way you get into the junk business is to have a friend, a father or some other relative in the game. The way he got in was probably by accident. He might have been out on a farm, and stuff began to pile up. Then he found out that he could sell junk for more than farm produce, and he shifted businesses. Or he ran a gas station, some cars got left around his place, and the same thing happened. Or he might have been a mechanic, and found out that other mechanics needed parts he could supply if he owned a couple of junkers.

But like any other complicated business, the fellows who know what they are doing make the money. There are maybe 15,000 parts in a modern car. Which ones are worth something, and which are just plain junk? Since there are maybe fifty different basic models turned out each year, the problem is straightforward—just know all about the markets for 750,000 things all the time, every day, for this year's models. One company makes money just printing an interchange manual, which is nothing more than a cross-indexing of all those parts. Your rear wheel bearing for a 1972 Ford just might be the same thing as the wheel bearing on an International pickup truck—and then again, it might not. If you know, it's worth money. The front left fenders of two different General Motors cars may look alike, but are they the same? Knowing for sure can be fifty dollars in your pocket. Or does one Ford engine interchange without fussing into some Mercury or not? There is plenty of gold in these hills for anyone with a good memory, the willingness to work a bit, and the ability to stick with it for a few years.

And if you don't like that end of it, you can have fun trying to buy copper, steel, or aluminum for remelting from the scrap-car people. Once again, you have to know what you are doing, but those who do end up rich. Given the steadily increasing American pressures to clean up the place, it is quite probable that the fellow or girl who knows what's going on can make his or her pile early and retire. But there is no easy, get-rich-quick way to do it. The best way to win is to get into the game (probably at low wages) and start learning. But then, what game *is* easy?

Planting Junk to get Rich Slowly

I have a friend out in Oregon who dropped out of school in the third grade. He worked hard all his life, retired at sixty-five in 1968, never having earned over $7,000 a year, and now sits on a $300,000 gold mine. His technique is workable for anyone who wants to get rich the slow way, and who really doesn't want to work too hard. This is what he did:

Back in the mid-1950s, this fellow decided to collect all the older junk cars he could find. He lives in a nice quiet town of 300 people, where there is lots of land along the river unsuitable for anything, so he got himself ten acres of it out of sight of the road. This cover was important, because even then people screamed loud and long at the sight of rusty hulks sitting around for years. He quickly found out that he could get more junk than he really could handle, mainly for nothing, so he began to specialize in Model A and Model T Fords, complete or not, in any shape or condition. By 1960, he had his ten acres full of rows and rows of these rusty hulks. Then he just sat for five years or so. As he said, "It's sort of like growing trees—you plant the things and watch them grow."

And grow they did. It turned out that old Fords became *the* collector's car, and lots of people began to restore them. When they needed a part, my friend would oblige, for a price. At first, business was slow, but the word got around. It may sound silly, but when a set of Model T wire wheels and hubs now sells for $100 to $150; or an undented 1931 roadster door goes for about the same price, it's kind of nice to have fifteen or twenty of them out in the back forty.

My friend invested perhaps a thousand dollars in cash and lots of spare time running around in his pickup truck and trailer getting the stuff. Last year, he was offered $300,000 cash for the lot, but he turned it down, in part because he can sell it off in bits and pieces for more than that, but mainly because it gives him something to do in his retirement. Lots of interesting guys come by to talk and trade these days, and while many oldsters are worrying about too small social-security checks and how to keep busy, my friend is worried about his income tax and how to find some spare time to go fishing.

It can be done again, and cheaply. Just do what he did. Find yourself an old, marginal farm, preferably up in the hills. You won't find any Model T Fords sitting around anymore (well, not too many), but off in any rural area are lots of abandoned 1957 Fords, 1954 Buicks, and 1961 Plymouths. Just haul them to your land and plant them. Make sure that you stick them off the road, down in the hollow, so that no one will yell at you. Best bets are convertibles, since antique buffs like them more than closed cars, but just about anything will do. If you feel like taking a flyer, try to figure out what people will want in twenty years (my guess—1946-56 Chevys, 1946-56 Fords, and Volkswagens—but you can take your choice). Then save just these. If you're right, you may have the only game in town in twenty years, and people will pay your prices. Since you can get the cars now for nothing except the cost and time of hauling them in, the rate of return on your investment can be substantial.

Anyone can play this game, and if you have ecological tendencies and like the open countryside, you can also help the rest of us out by concentrating the hulks in one place. If you're wrong, and no one wants the parts in twenty years, you can always sell them for scrap and make a few bucks. The way we are using up natural resources, it might pay just to have some scrap iron and steel around in twenty years. The price will be up anyhow.

Here and there around the country, and always way out in the boondocks someplace, are fellows like my friend who planted their junk cars long ago and sat. Now they are eagerly sought out by collectors, who happily pay nice prices to get that key part for their antique. It worked long ago and now, and it will probably work in the future. Play the game if you want.

City folks can't play that game, because it can take lots of space. But they can, if they care, have hundred-dollar cars, instead of the three-thousand-dollar ones most of us like. It's really easy, if you're willing to learn a bit about cars.

When you remember that the only difference between a new car and an old one is a small amount of worn metal here and there, the idea becomes clear. Just buy an old thing that's running—and there are lots of these around. Then find out how to keep it running with junk parts, as we suggested in the last chapter. Even if you don't drop out, it's a good idea.

A key part of this game is buying the car. Another friend of mine who does this notes that it is possible, if you know how to check them out, to buy a very good ten-year-old car for under a hundred bucks. He looks for cars that aren't running—often there is some minor thing wrong that he can fix. But he knows how to figure out what the problem is. Like everything else in junk, this is complicated, but learnable in a surprisingly short time. And he notes with pride that he is also playing his part in the ecological campaign. After all, every car that keeps on running means less new steel, copper, brass, and what not has to be produced. And somehow he always seems to have lots more money than most people he runs around with. They seem to always be paying for a new or almost new car, which really doesn't run any better than the ones he goes around with.

Recycling Cars the Right Way

If you have a half-million dollars or so, and you're looking for a business to get into, the idea of systematic recycling of older cars may be about to happen. This really is the antique restoring idea along a modified mass-production basis. Find yourself (from the junkyards, of course) maybe 500 1964 Ford Falcons, in any shape at all. Then recycle them, by taking them completely apart, reassembling them with new and rebuilt components, and sell them to ecologically conscious people, like college professors and other assorted intellectuals. We figured out that on a 300 per month basis, this could probably be done at a cost that would enable the recycler to sell the things at retail with a new-car guarantee for perhaps $1,500. The beauty of the idea is that you don't have to invest a billion dollars or so in dealerships, since Ford is still doing well, and if you recycle their cars, their dealers can maintain them. Hence the big investment in parts, shops, and all the rest is unnecessary. And your dealers could be part-timers, like college students in university towns, where such a car could be sold.

Most people right now would never touch a car like this, but recycling doesn't require producing millions of cars per year to keep costs down. You could make 300 a month only, and just possibly there are that many customers around who would buy the things. And for those who are ecologically concerned, this would be a good deal. If all you want is wheels, why bother with taking the trouble to make all those new parts to make a new car? Just make the old ones as good as new and start over.

This is already done all over the place. The antique-auto buffs do it, as do many bus companies. You can take a fifty-passenger 1958 GMC bus with perhaps 800,000 miles on it, strip it down, rebuild the whole thing, and come out with a like-new vehicle for maybe $4,000. A new one costs over $25,000, so if your customers don't mind, why not? Indiana University now does this with its fleet, as do many others. And trucks also are recycled, often in foreign countries, where new equipment is just too expensive. But it could be done here too, and some in fact do it.

This sort of thing is not very widely known, since, as I said, those who work with junk have no status at all. No one knows or cares what they are doing, except maybe some income-tax investigators. The idea that some greasy character working out of a big, messy lot with piles of cars around in various states of disrepair could be a socially useful guy has escaped us for a long time. But it is very likely that such activities will grow. Someone is going to have to take those junkers to get rid of them, and if new-car prices keep rising, it may yet pay to recycle the old ones. Moreover, more cars made each year means that more will be junked later.

We could change auto junk patterns fairly easily by insisting that cars were designed to be taken apart rather easily. So far, no one in designing a car gives any thought at all to how tough it is to separate all those things when the car finally gets junked. The relationship between the beginning, when the car is put together, and the end, when it gets taken apart, is exactly zero. No one ever thought about it. But it would be fairly easy to design a new car so that it could be taken apart easily—and

Making Money on Junk: Autos / 99

if it were, our junkmen would do just that. Ergo, no more unsightly messes around. Remember that there is lots of good scrap in any old chassis, but now it takes too much labor in a high-labor-cost society to bother.

We probably won't bother with this idea soon, because in the end, we are totally new oriented. I've personally had fun with young students by suggesting that they get into the junk business. They look at me as if I've finally gone totally over the hill. Prostitution sounds better—at least it's a respectable occupation. Drug-pushing is nasty, but it's nothing ugly, like messing around with good, clean junk. No, we'll leave the junk to the odd characters who backed into the game by accident, who have no status, no standing in the community. Which, if you want to make some real money, is just great. While all your friends are struggling along with their production ideas, trying to be respectable, you can sneak in the back door and corner a small piece of the junk business. No one will know or care, but you can laugh all the way to the bank if you're good.

5

Nothing Gets Used

Way out along some Arabian desert track we once traveled sits a 1942 Austin British Army truck. I last saw it in 1961, buried up to its bed in drifting sand. We stopped once to look it over, and as far as we could tell, nothing very important was wrong with it. Maybe the carburetor got full of sand; maybe the fuel pump failed. But the vehicle had sat there for ten years or more, untouched by anyone. Nothing was missing, except a few wooden floorboards. Probably some nomads had taken them for fuel in this treeless desert. But the tires still held air, and we figured that we could get it going again, if we had a bit of time and a few parts, within a day. We never did get around to coming back and getting the truck, since it wasn't worth the trouble. And for all I know, the thing may be sitting there still, a piece of virgin junk in an area where nothing much used to get used.

We have been wandering around the affluent United States, where sometimes the country seems buried in its own garbage. As we have seen, this is not quite true, since advanced countries also have very sophisticated junk/garbage-pickup systems. These may not be perfect, but at least most of the junk does eventually get recycled, one way or the other.

There is another kind of country or culture, however, where nothing really gets used. At least so little gets used that huge piles of junk just sit, often for decades or longer. Relatively few such places still exist, but they have their own particular fascination for a junk reader or user, as we shall see.

Such cultures are extremely primitive. They operate at the bare subsistence level all the time, using only human energy and perhaps burning wood in fires. This is the sort of culture that occasionally appears in *The National Geographic*, with titles like "The Lost Tribes of . . ." (fill in your own location). Because virtually everyone these days is in the economic development/nation building phase, such places necessarily have to be isolated. We can still find examples on remote South Pacific Islands, among the Australian Bushmen, and with the African Pygmies. Although such cultures may use a few modern artifacts, such as cotton cloth and steel axes, their technological level is Stone Age. Of the modern world, they know virtually nothing, and they probably have little interest in knowing much more. Our contact with such peoples comes by accident, since they have little or nothing to offer modern civilization. If they did, long ago the tribe was either destroyed, pushed off their lands, or moved into the mainstream.

World Wars and Junk

From time to time, the modern world does hit such cultures, usually by accident. One major example was during World War II, when both the Japanese and American forces occupied various subtropic islands for a few years. The foreign, modern forces moved in, brought with them vast amounts of modern military gear, then moved on, without ever bothering to tell the natives what it was all about. And when they moved on, the military junk was left on the islands.

Another example was when various Middle Eastern and European interests built the Hejaz Railway on the Arabian Peninsula before World War I. The idea here was to open up communications and transportation to Mecca, the Holy City of the Moslems, and the Railway (with connections) reached from the Mediterranean to Medina. Most of the desert part of this railway was built through isolated country inhabited by nomads with the same kinds of extremely primitive technology discussed above.

Then in World War I, Lawrence of Arabia destroyed the Hejaz Railway for military purposes. Bridges were blown up, trains derailed, and the usual pattern of destruction carried out. But unlike similar experiences in other parts of the world, the Hejaz Railway was isolated, in a part of the world subject to extreme political instability for a long time, and surrounded only by extremely primitive natives. And nothing happened. The natives quickly took anything they could use—mainly wood from the cars and locomotives. But locomotives and cars are largely iron and steel, and this type of culture cannot use chunks of metal that have to be melted and reformed. So the trains just sat, unused. They sat for over forty years, until the Saudi Arabian government finally rebuilt the line in the 1960s.

Railroad museums would have had a ball on the Hejaz. There sat perfectly preserved Krupp steam engines from 1910, unused, waiting. The gap between the technology they represented and the local culture was just too broad for them to be utilized. And the nice dry desert climate preserved perfectly the locomotives and cars.

On those tropic islands, particularly the ones occupied by Americans, the piles of junk were even more impressive. Try to imagine how a local citizen felt, when, one day, thousands of men came ashore, from ships bigger than anyone could have imagined, bringing such things as trucks and bulldozers, which ran in some mysterious way that no local person could figure out. In some cases, these island cultures had not even got to the wheel, let alone internal-combustion engines. Remember that on such islands there are virtually no metals, and distances are short. Without materials or incentives, invention comes slowly.

Moreover, the visitors brought an incredible variety of canned and preserved foods, chewing gum, metals in incredible abundance, clothing, airplanes, and all the other baggage of an advanced civilization. Everyone ran around doing mysterious things, like talking magically on radios or telephones. Their goals seemed very mysterious—what the hell were they up to?

And then, in a matter of months, everyone picked up and left. The remains of the invasion were left behind, including buildings, airplanes, trucks, and piles of stores. It wasn't worth it to bring the stuff back home.

Here was junk generation on a massive scale. What happened to it? Nothing much—it sat, and often still sits, out on those lonely atolls, sometimes half buried in jungle. From time to time some American or Australian will come to pick up some scrap, but the natives never will. They used up the obvious stuff, like odd clothing and whatever food supplies were left. But no primitive culture knows quite what to do with an almost complete P-40 aircraft, or a Zero fighter with lots of bullet holes in it. Even reworking the aluminum skin plates is often beyond the capabilities of the natives.

Find Yourself an Antique on a Tropic Island

Once in a while someone wants a museum piece, and he goes back to the scenes of major World War II action. He can find almost complete examples of aircraft for his museum which are unknown anywhere else in the world. Someplace in the Solomon Islands, as recently as 1972, sat the remains of a Japanese Zero fighter that took part in the attack on Pearl Harbor in 1941; scattered around these islands, among others, are examples of virtually every aircraft that fought in the early part of World War II on both sides. You can find machine shops, runways, buildings, and all the rest, slowly decaying. The decay rate is surprisingly slow for many items—the military built better than it knew at the time. In spite of hostile jungle climates, much of the equipment is in surprisingly good condition. A Japanese Betty bomber waits with its tire still holding air, after thirty years.

The years have taken their toll of this military junk, but according to Bruce Adams (in "Ghostly Boneyards of the South Pacific," *Air Classics*, August 1972), most of the junk is still quite recognizable for what it was, and one could probably build a complete Japanese bomber with the pieces that remain. The natives have only managed to hack out an occasional aluminum plate from the planes—most of them were abandoned because of battle damage, and most have lots of bullet holes in them. But scattered throughout the Solomon Islands, at all the old Japanese and American bases, are not only aircraft but machine tools, engine jigs, old truck tankers, and all the rest of the paraphernalia of modern war. Many of these aircraft are now totally immersed in jungle; others are stuck in the sandy beaches, or even visible under water in the shallow lagoons. Where the airstrips are still used for local service airlines or crop dusting, nothing much remains, since souvenir hunters and antique collectors have carted off the equipment. But if you want to cut through jungle trails for miles, you can still find all sorts of equipment. Even the scattered remains of the Betty bomber in which Admiral Yamamoto was killed in 1943 still lie peacefully on Bougainville Island, undisturbed after thirty yeears.

All of this is in the Solomon Islands group, where Mr. Adams visited. Now, if you really want some good stuff, read a Pacific War history, and travel to all the remote islands where troops gathered with their equipment. If the natives are at the nothing-gets-used stage, all the stuff will still be there for you to take, look at, or simply admire.

The armed forces of the belligerents went home (with a few exceptions: remember the occasional Japanese soldier who still gets found living on some remote island). The local natives remained, still not quite understanding what had happened.

Then a decade or more later, anthropologists noted an odd phenomena occurring here and there, called the cargo cult. A local wise man would see a vision of the coming of the cargoes again. The natives would begin to build crude airstrips or docks out of local materials to receive the cargoes. Once again, invaders from the moon, or wherever, would return with all their goodies—food, clothing, and all the rest—to raise the standard of living of the local people in some God-given way.

After a while, when the cargoes did not come, the island would gradually drift back to where it had been for a millenia or more, just a very primitive culture adrift on a course far removed from the mainsteam world, where junk not only gets generated but maybe even used. The gap between the cultures was far too wide to allow those at the bottom to climb out, and the contacts between the advanced and the primitive were too sporadic for proper learning to take place.

American Indians and Junk

These examples of unused junk have only begun to happen in the twentieth century, since before the mechanical world began, even very primitive cultures were close enough to the advanced to pick up and use innovations they saw. American Indians eagerly took to the horse, as one example, in the eighteenth century, when horses brought by Spaniards to the New World escaped. A culture used to four-legged beasts could easily understand how a horse could be used. The results were a very rapid increase in the Plains Indians' standard of living (did you ever try to catch a bison on *foot*?) and rapid culture change.

A major reason for the Indians' inability to resist the white man stemmed from their difficulty with guns, ammunition, and powder. The Indians could learn to use guns, but they had real difficulty with such trivia as cleaning them properly so that they would work right the next time, and with rifle calibers. Machine-age man understands quite well that a .30 cartridge will not fit a .41 barrel, but the premechanical man finds this all very mysterious. And when your rifle blows up in your face because you forgot to clean it last time and the barrel is full of crud, it becomes difficult to fight properly.

The ascending technology curve the white man was on would have doomed the Indian in any case. If the white men had gone away and left their railroads in 1880, these would have sat much like the Hejaz, for a long, long time. Preindustrial man could not have figured out what to do with them.

Everyone Learns in the End

Some of these unused items rusting on tropic islands have made some minor contribution to present affairs. You may own a Toyota, and, like most owners, you figure it's about the best car around. But where did Toyota get its start? Well, there was a Korean War in 1950, and suddenly, five years after World War II, the American Army needed lots of trucks and jeeps in the Far East in a hurry. Someone remembered those islands, where thousands of such vehicles had been left in 1945. The Army went back and picked them up, sent them to Japan, and had them recycled there. Then they were sent to Korea where they met the fate of most combat equipment—they got torn and shot up.

In 1959, I was looking over an order for a fleet of Toyota jeeps, which were new then, out in the Saudi Arabian-Kuwait Neutral Zone, far from virtually anything but oil. As usual, although the vehicles were brand-new, a few of them didn't work, and as usual, the spare parts wouldn't be shipped for some months. We took a look under the hood, and something down their jogged my memory. "Are there any old Chevy or GMC truck parts around?" I asked.

"Sure, plenty," one of the old-timers commented. "There used to be lots of them around, and many parts weren't used." We found a carburetor and a starter motor from a pre-1954 Chevy and tried them. They fit on the Toyota as if they had been made for it. Our jeeps ran better than anyone else's for a long time, because Toyota had apparently learned its lessons well. You see, those 1943 U.S. Army 4 x 4 trucks had been built by GMC, and Toyota had been one of the firms that had refurbished them in Japan for the Korean War. This got Toyota into the truck/auto business in a very big way, and when they built their own vehicles, they initially relied on a few things they learned from all that tropic island junk. It's really a small world.

Nothing Gets Used Until It's an Antique

For antique buffs, airplane collectors, and old-car fans, the thought of finding something just sitting there after thirty or forty years is enough to make one's mouth water. And as affluent Americans and West Europeans (and Japanese, too) begin to look for very common artifacts in their recent past, they often find that they have disappeared entirely—except on those sterile deserts or tropic isles. Then someone organizes a trip and, with a bit of luck, comes back with some almost complete remnant of times long gone, almost forgotten.

One fellow went up to Northern Greenland, at considerable expense, to haul home by helicopter a complete 1928 airplane, which had run out of gas trying to fly across the Pole. There was no one there, but even if some Eskimos had passed by, it is unlikely that they would have known what to do with an odd collection of steel wire, fabric, struts, and steel. In that climate, the aircraft was almost perfectly preserved—deep frozen for over forty years. And as war relics get more and more valuable, you might ponder a quick trip to Guadalcanal or Rabul Island to pick up what's left. Hurry though, since lots of collectors have now discovered that the stuff is there.

And there is a yarn about the Saudi Arabian desert, which may or may not be true. It seems that one old Arabologist, Mr. St. John Philby, was the Ford agent in Jedda for a few years. He sold several hundred Model A Ford touring cars around 1930 to the King of Arabia. They were duly shipped from Detroit in large wooden packing crates. But political troubles came, wars happened, and the king was occupied elsewhere. The crates were brought ashore and piled up inland. In that part of the world, sand storms regularly bury things left alone, and of course at that time the natives not only would not touch royal property, but probably could not have done much with the cars even if they had wanted to. So the Model A's were buried in sand—and they still are.

Now, if you have any connections with the government of Saudi Arabia, and if you could just possibly get a permit to go treasure-hunting, you just might find, outside of Jedda, all those Model A's in their original crates. Since this is close to Egypt, where the climate is such that the paint of King Tut's 2,500-year-old tomb hasn't worn off yet, it is very possible that the cars will be absolutely mint. With Model A touring cars going for $3,000 in almost junk condition, one wonders what two or three hundred brand-new, zero-mileage mint ones would bring.

Fast Learners

One reason that this sort of nothing-get-used situation is so rare is that local people learn very fast. If they stay in contact with others over long periods of time, they quickly figure out how to use things up, and we move to the next phase, to be discussed in the next chapter. Hence the Saudis, who are very quick learners, began to move out of the nothing-gets-used phase into the everything-gets-used stage within a decade after oil was discovered (1935). Only in a few far-off corners of the desert did the old order remain for very long. As soon as local citizens learned about trucks and cranes, the scrap iron began to disappear, to be sold for scrap at first but later to be used locally. The oil barrels, timber, sheet iron, old cars, and all the rest began to disappear, slowly at first, and then very rapidly. It is only in the far-off islands, deserts, and other very isolated places that primitive cultures can live out their destinies in peace.

If you want to drop out completely and tropic islands appeal to you, try to find one where American troops were stationed for a few years and where the local tribe is very primitive. Maybe, with luck, you can find some old aircraft to sell to collectors, and maybe there are still some canned goods stashed away in the palm trees to break the monotonous diet of coconuts. When you need to get off the island to have a gall-bladder operation or something, then you'll have the cash to make the trip. You see, these very primitive places always fascinate us, but what they really mean, when you see all that unused junk piled up here and there, is that people living there, including you, are going to be at the very lowest subsistence level there is. And this means no shots, no sanitation, no clothing, not much of anything, except the minimal food necessary to sustain life. If you get to such a place, be prepared to import everything, including labor, since no one around is going to be able to do anything.

Yet such places where nothing gets used fascinate us all. These are the ends of the earth, the places untouched by modern technology or anything else. These are the places where a person can go to get away from it all, really away, where nothing modern has ever been except by some odd accident. My mind goes back once in a while to that British Army truck in the Arabian desert—it was only just that it should break down and sit there for decades, because it really had no business being there anyhow. The desert was clean and pure, and far away from everything in this modern world. Hopefully, it will stay that way for decades to come.

But progress of a sort inevitably seems to come, and I suppose that some eager merchant or junkman has pulled the truck in. What happens is that countries and people get smart, and as they do, these odd places where nothing gets used drift quite rapidly into the situation where everything gets used. And that is worth another chapter.

6

Everything Gets Used

For a brief period when a non-Western culture first relates to the industrialized world, we find the nothing-gets-used junk problem discussed in the last chapter. But quite quickly people and cultures learn to use things, and we observe that poorer countries rapidly move into the everything-gets-used phase of junk.

This phase applies to most of the world these days, although few Americans, observing their own piles of junk, are likely to believe this. But some two-thirds of the world's population is very poor, with per-capita annual incomes below the $200 mark, as compared to perhaps $4,200 per capita for each American, or over $2,500 per year for Europeans. When people are poor, they normally don't generate much junk, and what they generate is quickly used up.

The two-thirds of the world we are talking about here would include all of Red China and India (right there about 1.3 billion people); plus most of Southeast Asia; all of Africa except the Union of South Africa; and well over half of South America. In these parts of the world, often called the Third World, or the Less Developed Countries (LDCs), most people have to struggle just to get enough to eat and obtain minimal shelter. There is precious little income left over for anything else. Moreover, this is the part of the world that is involved in the population explosion, where population increases from 2 to 3.5

percent per year. At these rates, population will double in 25 to 40 years, as compared to 70 to 150 years in more affluent countries. There will be many more people to share the poverty in a generation or so.

LDC Socio-Economic Characteristics

The LDCs have economic and social characteristics quite different from the United States, and as a result, their junk patterns are also very different. Following are some of the important differences:

Low-income per capita means that people don't make much money. A wage that Americans, even poor Americans, would find incredibly low would be seen as a good deal by most persons in such countries. Ten cents an hour would seem princely. If the average income for the whole country is only $110 a year (as in India), either few are working, or the typical wage is very, very low. Hence it is possible to hire people to do things that cannot be done in the United States.

Moreover, there are lots of people around ready to work. Unemployment rates in the poorer countries are often as high as 25 or 30 percent, just like the Great Depression in the United States. This stems from the difficulty of finding capital equipment to put people to work on—it is hard to finance such machinery when incomes are so low. It also stems from the population explosion—if many new young people come on the labor market, how can anyone find enough for them to do? And it is also a condition in agrarian societies, where farm workers only have a little to do during slack periods. But one key characteristic of any poor country is the battalions of bodies, typically quite unskilled, that are available for any purpose.

The LDCs also have very low levels of energy use per capita, which is another way of saying that they don't pollute things very much. Being poor, they do not have the huge, capital-intensive steel mills, paper mills, petrochemical plants, or even electricity-generating stations in large quantities. Items considered so common as to be casually tossed away in the United States, such as cans, bottles, and plastic packages, are so scarce as to be almost nonexistent in the poorer countries.

Since capital is very scarce, and labor is very cheap, the situation is exactly reversed as compared to richer countries. Managers and administrators use lots of labor if they can, along with very little capital.

Rapidly growing populations, improvements in agricultural technology, and some modest improvements in transportation and communications (like using trucks) have led to very rapid urbanization in most LDCs. Cities few Americans have ever heard of have millions of inhabitants, and most of the cities of over 1 million population these days are not in the affluent countries, but in places like India, China, Pakistan, and Bangladesh. These cities are strange for Americans in another way—they are inside out. That is, the poor live on the city outskirts in shacks, tents, or whatever else can be scrounged up, while the affluent live in the inner city where the American ghettos would be. The reason is that in poor countries there are very few cars, and, in such cities, living close in is a real advantage. You can see the same phenomena in any older American city—just look at those big Victorian mansions of the wealthy, built before the auto age. Many of them are at the edge of downtown.

In the modern LDC, we don't see Victorian mansions, but in the same location we do find high-rise luxury and middle-class apartments. Like older American patterns, a carless society is a high-density place, where lots of people per square mile get packed in.

Three Cultures, All at Once

It is also very typical to find two, and sometimes even three cultures all going on at once in the LDCs. The modern part of society usually has quite good incomes by local standards—say, $500 to $2,000 per year. These are the modern technicians, managers, administrators, entrepreneurs, and professionals, who inhabit those high rises. They are joined, at a somewhat lower level, by blue-collar skilled factory personnel, clerks, and other petty functionaries making considerably less money. All of this group will be less than 10 to 20 percent of the total population.

The second main group are the peasants. These people live out in the country, following traditions as old as time. Gradually they are being displaced by modern farmers with tractors, fertilizers, irrigation, and genetically selected seeds, but any poor country will have very many of these persons, up to 80 or even 90 percent of the population. This group is the poor one, often with cash incomes of under $50 per year.

And increasingly we discover a third group, which is in transition between the old and the new. These are the laborers and wanderers, who come to the rapidly expanding cities seeking their fortunes. They also have very low incomes, but occasionally someone gets lucky, finds a good job, starts a small shop, or gets to school, and eventually gets into the growing modern sector. Incomes here also are very low, typically under $100 per year. This means that such taken-for-granted items in the United States as clean clothes, a decent house (like a one-room place with a bath), and consumer durables like radios and TVs tend to be impossible dreams. Such people survive somehow in shacks at the edge of town.

As might be expected in such a situation, everything is in short supply, including medical care, education, and other people-improving services. Many adults (often up to 80 percent, but rarely below 30 percent) are illiterate, and many people are sick all the time. It is a grim world, but a lively one, since in virtually every such country, great efforts are being made to get into the modern world. Economic planning abounds; countries somehow manage to finance a few of the big new power plants and steel mills. These are never enough to go around, but at least something is under way. Roads are built, the prime minister promises that maybe next year every school child will get into school (they don't, but somehow more do get in than last year), and international experts pass through constantly looking for new ways to resolve poverty problems as old as man.

LDC Junk

Now, in this kind of world, what junk can you find to read? To begin with, there is precious little to find—everything does get used up very rapidly. Also, given the lower consumption levels, there isn't anywhere near as much around as we would find in the United States. No one provides shopping bags, and newspaper consumption in a largely illiterate society is small. Paper in general is hard to come by, and whatever is around gets reused immediately, for packing and writing. Wood, even small scraps, is precious in a world where any kind of raw material is expensive, and I have seen quite elaborate and well-designed houses built out of bits and pieces of wood that any self-respecting American carpenter would immediately throw away. But remember that labor is very cheap, while capital is very expensive.

Housing Projects, LDC Style

We once imported some autos into Arabia in huge wooden packing crates, about 8 feet tall by 8 feet wide by 12 feet long. We hadn't even begun to unpack them when a merchant had called on me to negotiate for the purchase of the crates. As any ordinary American might do, I was inclined to give them away just to get rid of them, but then before I could even answer, another merchant arrived to bid against the first.

Sensing that this was a bit different from home, we quickly

organized an auction, and forty cases were sold for around $40 each. As soon as we got the parts out of the crates, the victorious bidder triumphantly hauled his cases away in beat-up trucks. Then a few weeks later, on the edge of the city, I saw a new subdivision, neatly laid out in the desert, consisting of those packing crates made into houses. Each crate was a house, and the buyer of the crates had carefully cut windows and doors in them and even put some old auto glass in the windows. There were no floors, just packed sand, and the houses had only one room. But everything is relative—for a man who had lived in a tent with his family, this was a major step up. The contractor also had arranged for a single water line to be put right in the middle of the forty houses, so the families could have, for the first time in their lives, all the running water they wanted.

I found out that the contractor had sold the houses for $300 each, complete with water rights. Buyers were sober working men with families, making perhaps $30 to $50 per month. In that desert climate, it didn't really matter that there was no central heating or even any heating at all. And, as I watched the kids bustle around the new suburb, I saw someone toss an empty can out alongside the house—and no one rushed to pick it up. It would be gone the next day, since there were still poorer neighborhoods down the road, but these new suburbanites were beginning to learn their junk lessons well.

That was ten years ago—I suppose that by now mountains of trash have piled up, since the Saudis are not only a lot richer than they used to be, but quick to pick up foreign ways as well. I wonder how the garbage collection is going?

The Invisible Cans

Cans, bottles, and other containers get scrounged as soon as they hit the ground in really poor countries, since no one buys a set of dinner dishes when he is earning under $100 per year. In the early phases, the cans are used as is; later, some ingenious local craftsman begins to solder handles on them and bend them into more functional shapes. We have a gadget in the United States that cuts an old beer bottle in two, which is used by arts-oriented people for fun. In other countries, these little things get used commercially (remember those low labor costs). A craftsman laboriously cuts up bottles to make glasses for sale.

Tin cans are very nice, because they really are thin steel coated with tin, which means that they can be cut up and soldered and hammered into other shapes with very crude tools. And even in very poor countries, someone occasionally uses a can of food or drink. Remember that these countries don't have elaborate refrigeration equipment in either stores or homes, so the whole frozen-food revolution is in the future. If you want something not grown locally, you have to buy it in a can. And as the affluent do, those cans get back into the local market and get used.

What for? Well, toys, for one thing, both for local use if crude, and for export if refined. The Japanese used to make neat little toy cars by stamping them out of beer cans they got from the American Army camps. You could turn them over and see what brand of beer the Army was using at the time—the toy lithographing on the outside (which had been the can's inside) was new, but the underside had been left alone, and piece of the brand name was there. In Northeast Brazil, which is a poor part of a relatively affluent country now, they make (or made—these things change fast) ingenious local toys out of sardine cans, with soup-can lids for wheels. And they also made candle holders and lamps (who has electricity in such a country?), cups, plates, and a wide variety of artifacts, all ingeniously put together so that the brand name of the sardines came out right on the outside of the finished products. Here they didn't have local paint, or money to buy any, so the can ornamentation had to do.

Big cans, like those number 10's used for commissaries

and restaurants, can even be used as a building material. I once saw an entire house sheathed in them, shining in the desert sun. It was not only cheap, but a useful idea, since the bright tin reflected the heat of the sun away from the house. (I just used a soup can yesterday to get a little piece of metal to fix a gismo in my car—the art is not lost, but few Americans bother anymore). But, if you live in a snow climate, and your five-year-old car is beginning to rust through in spots, try getting one of those big cans, like a two-gallon oil can, and cut it up. Using small nuts and bolts to hold the tin, cover the rusted out areas, then smooth it off with plastic body putty you can buy at any auto-parts store. A few hours work can add a hundred dollars of value to your car. Yes, tin cans are one of civilization's great inventions.

Toilet Paper, Music, and Other Necessities

Most Americans would die without toilet paper, but there is precious little of it around poor countries. So, here is one major use for old paper. When I was in the Middle East, I noticed that my English *Economist* magazine was always disappearing immediately. It was the airmail edition, done on ultralight paper. We discovered later that this excellent magazine made better toilet paper than any of the local commercial brands on the market, so we guarded old issues carefully. But it's all history now—the magazine has changed its paper stock. And, given the small supply of paper and the large number of people, most citizens have to get by with old newspapers or sand—the traditional paper.

Most of us are familiar by now with the oil-drum bands of the West Indies. Wherever there is oil, there are oil drums, and they also show up in various places where major contracting work is being done. These big, heavy 42-gallon drums have all sorts of good uses, the most common one being for housing. You cut the heads off, then flatten the barrel, and what is left is a nice sheet of heavy-gauge steel, which can be used for good home construction or roofs. You will see such houses wherever the drums can be found, and there is always a lively free market in the local economy for such barrels.

Making musical instruments with such things is also a part of any poor country's life. A few older Americans may even remember the cigar-box violins made out in the hills, or even Bob Burns' Bazooka, made out of some old plumbing pipe; similar instruments are found here and there around the world, including the steel-drum percussions. Bottles, cans, wood boxes, steel tubes, copper pipe, and Lord knows what else get plugged into the local culture. Results range from discordance to some very lively and creative music, which many Americans have not discovered yet. They will though—various countries have ingenious local craftsmen and musicians now at work on sounds, mostly for their own amusement.

Even really crummy junk like broken bottles get used. The rich in poor countries have a lot to lose, and they tend to live behind high masonry or stone walls. On top of these walls one often finds broken glass embedded in concrete, to discourage climbers. Even old chunks of broken concrete and rock get used—here, local contractors make up walls and roads, using such materials as a filler. Concrete seems to be one of the first industrial materials available in most poor countries, and it is surprising what can be done with some concrete and a lot of broken rubble.

The Autoless Society

Since there are very few autos, and those that are around tend to be very expensive (in many countries, an ordinary new American sedan might cost over $20,000, because of overvalued local currencies and very high tariffs), one rarely finds an auto junkyard. Citizens of poorer countries would view a typical American junkyard with total astonishment. The United States scraps something like three times as many cars every year as exist in all of India. By Indian standards, most of those scrapped would hardly have been broken in yet. Indeed, if an Indian entrepreneur could get his hands on the inventory of any typical American junkyard, he would be fantastically wealthy. For one thing, most of the cars in the yard would be running again within a few months—just because countries are poor, does not mean that they do not have some very good mechanics around.

I saw a 1931 Nash being used as a taxicab in Alexandria, Egypt, in 1961, and I suppose that it is still running, with well over a million miles on its odometer. Model "A" Fords made in 1930 still serve as taxis in Lima, Peru. What happens when labor is cheap, cars are expensive, and people are poor, is that a large number of local repair shops spring up, which not only fix cars but make parts, too. You can wander around Cairo in the back streets and, if you have samples, literally have any existing car built for you. And it wouldn't cost much more than the vehicle did new. Egyptian craftsmen can do anything from making a radiator to manufacturing a clutch bearing, just by looking at the old part. But watch the quality control, particularly for moving parts. Not every indigenous machinist knows all about heat-treating and other metallurgic mysteries.

What to Do with Your 1926 Essex

In LDCs, when a very old (say, 1926) car finally comes to the end of the road, it goes off to a small scrap yard. There it is taken apart, down to the last nut and bolt and piece of wire. With labor at ten cents an hour and bolts at five cents apiece, it does not require too much work to get a day's wages, along with a profit for the lucky owner of the junker. And then the parts get reused. Nuts and bolts are salable for many purposes; the plate glass in the 1926 Essex can become window glass in that oil-drum hut; the copper wire will find its way into the electrical market and radio repairs; the steel sheets on the roof and doors become valuable raw materials for some local factory; the tires get cut apart, both for the wire in the beads, which can be sold, plus the tread, which goes to the local sandal shop; the cast iron in the engine, laboriously segregated from everything else, will be sold to the local foundry, which casts drain pipes for the big new high-rise buildings; the seats may end up in a barber shop as is, or might be taken to a local furniture dealer for reupholstering and rebuilding into more modern Western-style furniture; and the copper tubing might end up as a part of a simple still (legal or otherwise) in some small shop. In the end, what is left is an absolutely stripped hulk of pure steel, and even this might find its way into small manufacturing plants. If the country has a fledgling steel mill, the hulk goes there for scrap; if not, sooner or later some Japanese or West German buyer comes along and buys a shipload of clean, good scrap to be sent back home.

Once in a while you may stumble across the one big scrap yard in the country, where these stripped hulks are being accumulated for shipment, but otherwise, there are no junkyards as we know them. Autos whose days are over resemble, after a very short time, dead animals in vulture country.

The Invisible System

One curious point about all of this is that nothing is written about it. All political leaders of poorer countries are convinced that the only way to fly is to be as Western as possible, as quickly as possible. Anything smacking of junk, residual garbage, and similar Western things is taboo for discussion in proper circles. It is very common to discover that the country's leaders and industrial planners do not even know about these little back shops manufacturing 1935 Ford pistons, or handmaking 1953 Buick radiators. They are so irrelevant as to escape the conventional types of industrial census.

Hence anyone who even suggests that such goings-on are a real asset to the country is dismissed immediately as a foolish eccentric trying to sabotage the best development efforts of the country. I know, because I've tried it on many occasions. The only thing that counts is new factories, new industries, heavy investment, skilled labor working in such big places, clear-eyed administrators in air-conditioned offices, and all the rest. This is fine, but unfortunately in most poor countries, these attitudes have the effect of making the few richer, while most poor stay poor.

This development follows from the vast numbers of really poor people around such LDCs, plus the population explosion. There is no feasible way that the world, or any one country, could build enough modern plants, schools, houses, and office buildings to fix up the poor of this world in less than five hundred years. Hence we are stuck with what we have, and that is, unfortunately, a bunch of semiliterate expert craftsmen on some back street of the capital city doing wonderful things with scrap iron, old nails, and assorted pieces of junk that no one knows or cares about.

The very different auto situation between poor and rich countries does suggest an intriguing way to raise the living standards of both the rich and poor. Remember, to a man making $100 a year, an old junk car would be a mountain of riches. To the affluent American, it's a painful piece of junk to get rid of as soon as possible. So why not match the need to the surfeit?

War 1980 Style

We might ponder the following strategy: instead of messing around killing peasants in dismal places like Vietnam, we should

Everything Gets Used / 123

declare war on India. This may seem extreme, but it's the only way to get the Indians to go along with what follows.

When war is declared, we bundle up a couple of million junk cars in chartered ships and send them to India. Since we scrap over 6 million cars a year, and since several million pile up unused, this is easy to do. If ordinary tramp steamers were chartered, it might cost $100 per car to get to India, but, after all, even one bomb from a B-52 costs $10,000 or more, so this would be cheap.

The cars would be dumped on various Indian beaches, far from cities, and the attack would be over. The Americans would go home. Americans, thinking about 3 million cars dumped on perfectly nice beaches, are appalled, thinking of all that contamination and pollution, but relax—within three months nothing will be there. What the typical Indian citizen would do is rush out and begin to stake claims. Before very long some (high) percentage of those so-called junkers would be running smoothly around Indian streets, probably as buses or taxis. The others would disappear in small pieces in the manner described above.

Lots of Indians would learn new skills, since messing around with old machinery is very educational, and there just isn't enough old machinery in India now to go around to all the young men and women who might want to learn. The perennial Indian transportation shortage would be alleviated somewhat, and a considerable number of Indians would be much better off than they were before.

On the American side, the federal government buying junkers would push the price up nicely, which would make it possible for our own to collect them for export as war goods. Many citizens, seeing that the price of really old cars was going up, might be encouraged to keep theirs just a year or so longer, reducing scrappage. People in big cities would happily get their junkers out to the sales yard, reducing abandonments. And steel scrap prices might even rise nicely, thus shifting demand into new iron ores and encouraging a bit of employment.

Brace yourself for the United Nations showdown, however. The Indians would be screaming about imperialistic aggression, fouling the environment, and all the rest. So, for that matter, would all the American liberals. The reason is simple—these people have all attended the same universities, and read the same books. And because they have, the idea of using junk in any size, shape, or form is totally alien to their own way of thinking.

But meanwhile, back in India, the government would discover to its surprise that it was becoming very popular with the common man. Such very scarce items as wire, glass, and steel are now available in abundance, and prices are down. Lots of ingenious poor Indians are now taxi drivers and they even own their cars. There would be, of course, a fuel shortage, and local government officials would be very busy figuring out how to get more gas, plus how to ration scarce supplies. It is likely that whole new markets for all the bits and pieces of autos now available would be created, involving still more efforts to tax and control. Things would be very lively. The local communists would be demanding that all the cars be shared equally among all Indians, rather than be allowed to fall into the hands of exploiters who happened to live near the beach, or got there first. One can easily imagine the turmoil, opportunities, confusion, and political maneuvering that would accompany all this new wealth.

And lots of Indians would be getting an education of a sort that they could not have had before. Even 3 million cars would more than double the number now in India, running or scrapped, and the craft industries would have to hire apprentices and workers in those dismal back-street shops to meet new demands. Since most of the cars would be old, there would be lots of work to get them running, and keep them running well, and no big organization in India could figure out how to do this in the short run.

A few clever entrepreneurs would go even further. India needs trucks and buses a lot more than cars, but trucks are big and hard to ship. So, with too many cars around, why not make trucks and buses out of them? This would involve body-rebuilding, chassis-lengthening, new springs, and all sorts of things that an ingenious technician could do, particularly if he could lay his hands on, say, 200 1963 Ford Galaxies at once. And, with 3 million cars involved, he probably could. Now all he has to do is work out his design, rebuild the key components such as engines and drive trains, and build up the rest. He might end up doing what Toyota did in the early 1950s with those U.S. Army 4 X 4's, and this would be a very useful thing for the Indian economy. Japan has done very well, after all.

If this war worked out reasonably well, we could attack next year with another 2 million junkers, and then again the year after. Or, just to be fair, we could declare war on other poorer countries. The Philippines offers mind-boggling opportunities, given their skills with rebuilding old jeeps into taxis and buses; Egypt is another nice choice; and Pakistan might also be let in on the game.

War as a Positive Sum Game

Too bad it will never happen. Wars are supposed to be losing games, but there is one where everyone could win, even if they didn't think so at the time. But the thought of Pentagon communiqués gravely announcing how many thousands of tons of junk got unloaded in a given day; the observations of what the enemy was doing with it; and all the rest, would be a welcome change from the usual grim war stuff.

One can see a lot of junk in countries where everything gets used, but it is not lying around. It is being carried someplace by a porter in a basket, or by an animal-drawn cart (made, incidentally, from an old auto frame—they learned it from us). Or it is being sold in the local bazaar, in bits and pieces. In lots of countries, a familiar sight is a street peddler, his total wares—consisting of a few old cans, some beer bottles, a couple of keys that don't fit anything, and similar stuff—spread out on a cloth in front of him. But no one is going to leave any junk lying around.

People and Junk

Helping out with the rapid disappearance of any junk generated is the way the population is placed. The junk generators are the big high-rise apartments; these are usually locked into the junk distribution set up by having the manager or concierge arrange with someone, often a relative, to scan all garbage for useful items. Since the relative lives a few miles away in a shack at the edge of town, it is easy for him to get in once a day to carry out the bottles, cans, and other assorted stuff. Organization of such markets by supposedly uneducated people would require a really good economist to trace. And, since no one pays any attention to junk, the market is always a free and uncontrolled one. Anyone can do anything at all and get away with it, if he can fight off his competition.

Here is a nice steady flow system. New stuff comes in at the port from abroad, or is produced locally and sent to the city center. Wealthier citizens buy it, use it, and discard it. The junk then flows back to the edge of town, where the poor live and use every scrap of everything. From time to time, some of the juicier items, like pieces of wood or larger cans, end up on some back street, where furniture or parts are made from it.

Total Recycling: What the LDC's Can Teach Us

In a few cities in poor countries you can discover a modern sewer system, but many parts of the city, and lots of provincial centers, don't have one. Sewers are capital-intensive and very expensive. So even human feces gets recycled. Honey-bucket carriers are known in Japan, but many other countries have them too. The feces is carried out to the fields for fertilizer, and the recycling process continues.

Note that it takes considerable skill and organization to continue recycling *all* junk. Most Americans would be unable to use the things that form a routine part of the total economic picture in a poor country. And here is where junk reading can be very useful to anyone interested in doing any industrial or economic activity in such a country. A walk around town just looking at what is going on can pay dividends. There are lots of skills and abilities hidden in junk processors and users in any poor country, but few people know how to find them. But just follow the instructions we mentioned in Chapter 1, and be sure to find those weird back-street shops where things are going on that even the most knowledgeable government technocrat knows nothing about. It usually turns out that there is a lot more in a country than anyone realizes, and all sorts of interesting skills abound. If you know where to look, you can find them.

As you might expect, there is not much of an ecological or pollution problem in the LDCs. Not only are few things used, but everything gets recycled right away. Persons from these countries who have not been in the affluent West very much find it puzzling that Westerners are so uptight about pollution. After all, in examining their world, they find very little. Only in rare cases where some big Western operation is going full-blast, like a petrochemical plant or oil refinery, does one find the sorts of really sticky pollution problems so common in the West. But these countries are learning fast, and they are beginning to develop new kinds of junk.

You see, this stage these days is just the prelude to bigger and better junk piles, along with more income, better health, more education, and other nice things in life. As countries get richer, their junk changes. But that is worth another chapter.

7

Piling It Up

I was driving on the edge of Beirut, Lebanon, years ago, when I noticed something new. It was an auto junkyard. A yard had been there for the three years I had been in town, but it had been a scruffy industrial junkyard for non-auto things of the poorer sorts. Lebanon at that time was not a wealthy country, and they were still pretty much in the everything-gets-used category. But from time to time I had dropped in to buy a chunk of pipe or a roll of copper wire.

But this time the place was different. Five ancient cars were sitting in a row at the edge of the property, and while bits and pieces had already started to disappear from them, they were still more or less intact. There had always been one auto junkyard nearer the town center, but as a wise old English mechanic once correctly told me, "There's nothing in it!"

True—by the time a carcass got to the yard, it had been well stripped. But this time—there they sat, forlorn old prewar Citroëns and Hillmans, with bald tires, ripped upholstery, and wrinkled fenders. I stopped and chatted with the owner. Yes, he had decided to buy a few cars, since they were now a bit cheaper than they had been. But he was disappointed—somehow, the parts weren't selling as he thought they would. And it really wasn't worth getting all the nuts and bolts off of them, given the high cost of labor. Why, one worker wanted the equivalent of two dollars a day, just to be a stripper! It was obscene.

I had been away from the United States for some years, and the sight of those junkers just sitting there was a welcome sight. Lebanon, growing steadily richer, was right on schedule.

A new factory had opened a few days before, and there had been lots of fanfare. Even the Prime Minister had come over, with his limousine and motorcycle escort, to clip the opening ribbon and make a speech about how rapidly the country was developing. Industrialization was the way to go, and this new factory, owned by a Lebanese group, was going to make aluminum extrusion products and some auto parts.

It had been a familiar scene, since Lebanon was doing lots of developmental projects then. Almost weekly we could read in the local newspapers about new accomplishments, new triumphs. But no newspaper covered the junkyard's shift to autos, and no one cared whether or not the crafty owner was going to make a pound or dollar or not. It really didn't matter, since the junkyard was in the worst part of town, and anyone who expected the Prime Minister, or even a junior government clerk, to climb around that dirty scabrous, rubble-strewn mess in the junkyard was confused. Clearly it didn't matter.

The new aluminum plant was in receivership within a year, and it has been having financial troubles ever since. The last time I checked the junkyard a while back, it was doing fine, and the owner's sons were now working with him in the business. In just a few short years, the yard had gone from five old junkers to over four hundred old wrecks, and had moved to a new location still farther out. Fellows who look in the back door instead of the front sometimes come out pretty well.

The Poor Do Get Richer

One characteristic of our modern world is that countries

and people tend to get richer through time. Anyone observing the present state of poverty and misery anywhere in the world may question this, but indeed countries grow, and people grow with them, at least economically. Those very poor countries where everything gets used gradually do build their factories, schools, power plants, roads, and what-not. Very slowly at first, but more rapidly later on, the economy tends to improve.

In the income bracket of from $400 to $800 per capita per year, Brazil, Mexico, Argentina, Taiwan, and Singapore are examples of countries that are a lot better off than the really poor countries. And Ireland, Spain, Portugal, Yugoslavia, Lebanon, and southern Italy are a bit ahead of these places, but not by too much. Such income levels are still less than a quarter of those in the United States, but at this level the really grim poverty associated with India and Bangladesh has partially been alleviated. As incomes rise, the composition of junk also changes, along with the way it gets handled.

High incomes per capita mean more pay for everyone, and, like people everywhere, most citizens spend most of it. They buy clothes, better housing, and better food to begin with, and because the production capabilities of the country are much better, such stuff is available. Packaged foods, more canned goods, and synthetic clothing get on the market in a major way.

Because a more skilled population is a more literate one, more newspapers get sold. In many cases, the first big paper producer gets established, so there is a lot more paper of all sorts around. In the better shops, and even in middle-class neighborhoods, you occasionally find your purchases being wrapped in brown paper instead of old newspapers. Paperback books get into a growth situation, as do magazines.

And the automobile age begins, slowly at first, but then often very rapidly. In 1957, Lebanon had around one car for every forty people; by 1965, there was one car for every ten people. One characteristic of these countries is what seems to be a perpetual traffic jam around and in cities, as far too many cars jockey for far too little space all the time. No one in world history has ever correctly forecast the coming of the auto age, and it is unlikely that anyone ever will.*

Labor is getting somewhat more expensive, particularly skilled labor, and lots of simple things that used to pay off, like cutting tires apart to get the wire, gradually become uneconomic. Typically there is still lots of unskilled labor around, often unemployed or semiemployed, but by the time a country gets to this level, it also has developed some strong ideas about welfare and minimum wages. The usual result is that unskilled men in from the country can't find work, since employers have to pay too much, given their productivity. And such things as picking up trash never seem to get done right because no one has enough money to afford all the men required.

These countries are also beginning to get into industrial pollution of all sorts, because now they do have the messy petrochemical plants, paper mills, coal-fired power plants, and all the rest of the industrial establishment that is noted for such pollution. In the early stages of the development, such pollution is typically ignored, since it is minor; but if industrial development is concentrated, and if atmospheric conditions are right (or wrong), then even a fairly poor country can get into fairly massive pollution problems rather early in the game. And typically there is no one around who really knows what to do about it.

Japan, which only ten or fifteen years ago was in this phase, knows all about this. You have to wear an oxygen mask in Tokyo on smoggy days, and cars are being banned from the city center in a desperate effort to keep people breathing. But such situations are the exception, since most countries in this

---

*One partial exception is the Sheikdom of Kuwait, which had great oil wealth when the country was still very poor, plus a ruler who had the good sense to hire a very excellent city planner very early in the game. This fact, plus the fact that the country was largely open desert to begin with, led to a city that vaguely resembles Los Angeles in layout—big, wide streets, lots and lots of cars, and all the rest. Esthetic it may not be, but practical it most decidedly is.

position are as yet too undeveloped industrially, and have too few motor vehicles, to get into really dangerous pollution conditions.

Washing Machines and Maids

In such countries, consumer durables are beginning to come in strong, too, since the price of maids and servants is going up fast. But market penetration is still low. It is common to find that perhaps 70 percent of families have radios; 20 percent, TV sets; 10 percent, washing machines; 2 percent, driers; 5 percent, vacuum cleaners, and so on. These figures rise rapidly year by year, since mass production of consumer durables is fairly straightforward, once the system is designed and organized.

When such things break, they get fixed. Such countries normally apply very high import duties to consumer durables, in part to encourage local production, and in part to serve as an income tax. Countries in this income category have real problems in applying an effective income tax, and most wealthy citizens try to evade whatever tax there is. But it is possible to police the few ports easily, and import duties can be levied on visible consumer goods. Since only the upper middle classes and the wealthy buy such things, the tariff is paid by those who would pay high income taxes in more developed economies.

Since production runs are low at the outset, costs are high, which makes all durables expensive, be they manufactured locally or imported. It is common to find refrigerators, TV sets radios, vacuum cleaners, gas stoves, and similar items selling in such places for two to five times what they cost in the United States, while incomes are much lower. When your vacuum cleaner breaks down, you don't throw it away, since a new one might cost $300 to $500. You can find a repairman who works for perhaps 80 cents to a dollar an hour to fix it.

One result is that there are some fascinating home consumer-durable antiques in most of these countries—things just keep getting fixed. And a byproduct of this high cost is that you never see a discarded appliance in a junk pile. It is a cleaner, less cluttered world in this sense.

Cars Get Made, Too

Cars follow a similar pattern. Countries like Mexico, Spain, and Brazil want their own auto industries, so they rig import tariffs to make sure that they have one. Then cars cost 150 to 200 percent more than they do in Western Europe and the United States. This means that you often fix instead of discard. And if you're rich enough to discard, the fellow who picks it up will fix and sell, not scrap.

In wandering around this type of country reading junk, you quickly perceive that this is a new situation. Really ugly junk begins to appear, often all over the place. Broken bottles, occasional piles of rusting cans, and sheets of paper blowing down busy streets are some of the signs, as are occasional piles of very odorous garbage, which somehow haven't been picked up for some time. By this time, these countries have got their sewers built, at least in the major cities, so human feces is not much of a visible problem, but public urinals abound and smell. And in the sweep of progress, it turns out that somehow no one really remembered to organize junk and garbage collection properly—there are so many things to do. As a result, the pile up begins.

But the better varieties of junk are nowhere to be seen. Old clothing is carefully collected to make rags; most paper is still piled up for use as wrappings or recycling; bottles are usually the deposit type, so someone has time to pick them up. Remember that in this type of world, wages run from 50 cents to perhaps $1.50 per hour for highly skilled men, and unskilled unemployment is a major problem, so anyone who can organize a good salvage operation is able to make some money.

The Golden Age

This is the golden period of the junkman. A smart entrepreneur can find dozens of men looking for work; he can organize his sources of supply, and arrange for resale of his salvaging operations with little difficulty. The price ratios between raw materials and labor are such that it pays to get organized in the junk business. It is hard to remember that around 1910 to 1930 the United States was in this period, and this is when the now big salvage operations got started, usually by Italians or some other ethnic minority. Clearly the WASP majority was not about to get into junk, so many profitable opportunities were left to others. And they took full advantage of it.

Similarly, in these not quite poor countries, some minority or another is quietly putting together junk empires, getting garbage contracts, and otherwise getting fixed up to make some big money. I have seen (as many other consumer-goods sellers have observed) a man who smelled a bit and looked as if he really didn't belong to nice society reach into his pocket and peel off $15,000 in cash for a new Mercedes. It turned out that he owned a salvage yard at the edge of the city. One of the delights of junk, up and down the country scale, is that no one in authority is paying any attention, so the possibilities of tax evasion are very high, and the probability of getting caught is very low. With a business that is growing daily as the country expands, along with no taxes to pay, it is really not too surprising that these smart junkmen often are pretty well off financially.

So we see the beginnings of salvage yards, recycling operations, and similar junk-related activities. Various kinds of industrial scrap are also expanding rapidly, so more options open up. Where metals are involved, one begins to see smelters, which melt down the steel, iron, brass, lead, zinc, aluminum, and so on, sometimes into pigs, often into usable new items, such as concrete reinforcing bars or cast-iron pipe. When you see such a junk operation in a supposedly poor country, you are really looking at the future. There has to be a lot of scrap in the country before such an operation can begin—apparently wealth is going up very fast. There has to be a rather sophisticated organization for obtaining the junk, transporting it to this place, getting it sorted out, and processing it—clearly managerial-skill levels are rising fast, and in an area unnoted in anyone's *Statistical Abstract*. And often the men operating the system are more skilled than anyone would expect in such a country.

The First Auto Graveyard

Auto and motorcycle junkyards begin to appear at this stage as well, but they are largely empty. That is, the cars and hulks there are very old, since autos last a long time in this situation. And almost every nut and bolt is gone. Labor costs are still low enough so that it pays to dismantle a vehicle completely before it is melted down for scrap. The back-street shops so common in the poorer countries begin to change their product mix in this more advanced country. Instead of making things, shops often now replace things. If your clutch burns out, there may be a usable spare in one of those junk yards. Or perhaps it will be remanufactured by replacing worn parts, not completely fabricated. Similarly, the various other major components of autos will begin to be remanufactured, such as engines, drive trains, and electrical parts. If your generator burns up, instead of having the shop take yours off and rewind it, they will put a previously rewound one on, taking your old one in exchange. Because some parts are now available, this is a logical and economic shift in the whole repair/replacement system.

Reading Junk in Construction

One easy way to see how far the country has gone with junk is to walk around the always booming center of town and observe the new high-rise buildings being constructed. These buildings almost always are constructed by local contractors and labor, so you can see how the locals are doing. Watch for these items:

Hardhats are always worn by someone, but are they worn by everyone in the crew? Since hardhats are expensive, they say something about the value of life in the country, as well as the skills of the workers. The higher the skill level and the pay rate of the laborers, the more interested the contractor is in their safety. No smart operator wants to lose a really good worker—it costs too much, if nothing else.

The scaffolding and forms used are very relevant. Modern countries use prefabricated steel forms, which can be bolted together to fit any site. If you see these, it means either that some foreigner is financing the project or it is a government job. Such things cost far too much for the local economy. The more common pattern is to find scaffolding made of odd bits and pieces of used wood—and when this job is done, the scaffolding will be torn down and carefully shipped to the next site.

If the cement trucks and similar equipment look like fugitives from postwar Detroit, this is about right. If not, the country is coming along even faster than you thought. Trucks tend to last even longer than cars, since they are much more expensive, and capital is very hard to come by. Of course they will be smaller and less productive than modern equipment, but remember that labor is still much, much cheaper.

Watch how wiring and plumbing is put in place. In these not quite poor countries, it is common to build everything on site—prefabricating is unusual. Again, labor is cheap, while capital is expensive, and it takes lots of money, factories, and skills to prefabricate a wiring system or cooling/heating system for a modern high-rise building.

Often the concrete is mixed on site in such places—watch how lots of men use small mixers, which usually are very beat up and old. The big, modern, 20-yard cement-mixer trucks are not around. And of course you will find that things get picked up fast around the site. If the workers cut off two feet of steel reinforcing bars on the top of a column, these ends will either be sold fast by the contractor to one of our new junkmen, or the workers will get away with them before the day is over. Even bits of wood and similar scrap will be gone quickly.

If you pass by after work, you can also determine a lot about the country's moral climate. Some sites have barbed-wire fences, armed guards, police dogs, and other signs of an uptight security system. Others, like Japan not too long ago, are open and unguarded, even though there are lots of goodies waiting to be taken. If a contractor can trust anyone who wanders by, it is very possible that any employer can trust people a lot further than he might expect in a relatively poor country.

You can also see, if you look closely, lots of salvage going on right at the sight. Workmen laboriously pull old bent nails out of pieces of wood; carpenters cut and saw to make new forms from old lumber; and electricians, helpers carefully segregate the bits and pieces of copper wire cut off of longer pieces that went into the building. This salvage material may go right back into the building, or it may go to one of those new junkmen who like nice segregated scrap. Either way, there is a lot more salvage and string-saving going on than most Americans would believe. It hasn't happened in the United States since 1940 or so, if indeed it happened then.

Sewers Carry Junk, Too

One important thing happens during this junk stage, which is the rapid organization of junkmen, both for cars and other junk. Junkmen, like other people, get accustomed to doing something well; as their world changes, they keep right on salvaging things, although not always in exactly the same way. But as more junk appears in the next stage (to be discussed in Chapter 9), someone is around to handle it. The problems to be solved at this earlier stage of modest affluence are great, and creative junkman learn to be both imaginative and flexible. This is fortunate for the rest of us, since if they did not, we would have been drowned in junk long ago.

Also in this stage we discover that the sewer problems are being solved rapidly. We tend to forget that only a hundred years or so ago, highly civilized places like London had open sewers; now, even modestly affluent countries begin sewer construction along with running-water systems, quite early. Hence the really smelly garbage and sewage begin to disappear, at least in the cities, in this phase, and we hear no more of them as consumer problems from now on. We also hear very little of typhus, typhoid, diphtheria, and other deadly diseases caused by improper sewer practice. We may be on our way to more junk, but filthy, disease-producing sewage and garbage are on their way out in this phase. Infant mortality rates in these slightly rich countries already are beginning their dramatic decline to levels common in the affluent countries. There is a world of difference between human sewage as junk and some nice clean copper wire. Count your few blessings.

We do hear lots about what to do with the sewage once it is collected and piped to some outlet, and countries at this level typically have many problems with cleaning up such messes. The reason is simple—virtually all sewage processing is capital-intensive, with very little labor content. After all, such processes were designed in Europe and the United States, where labor is expensive and capital abundant. Poorer countries have troubles getting enough funds and equipment to do the job right. You can also smell this problem from time to time, when you go five or ten miles downstream from the city and just stand near the river. Raw sewage is just dumped in. Or the ocean for miles around the port has few or no fish—the offal has driven them out. But even very rich countries now have similar problems. Slowly but surely, the processing plants get built, and this messy problem becomes a minor one, the province of capable technicians.

As the country develops and expands, however, its garbage-collection system gets into deep trouble. Remember that really poor countries don't have cans or bottles in their minimal garbage—these get removed before the garbageman ever arrives. But slightly richer countries do begin to find such things in garbage pails, and the tonnage of garbage collected per year begins to rise geometrically. Suddenly, the always inadequate garbage-dump sites are full, and new and much larger sites have to be found. This is, so far, a never-ending process. You no sooner find a good, nearby site than it fills up five or ten years faster than you expected, so you have to go find another one. It would be fun to find out how many cities and urban areas of over a million people are right now looking for a place to dump garbage—I would guess most of them are.

While pundits try to forecast gross national product, industrial production, housing starts, interest rates, and all that, no one ever sits down and forecasts garbage and junk flows. These things simply happen, and in countries that are beginning to get rich rapidly, as are the ones discussed in this chapter, the incredibly rapid expansion of junk, garbage, and sewage catches everyone by surprise. You can see this point easily enough by going back to 1935 and reading up on American problems. Try finding anything about junk. A bit about sewers as a health problem, maybe, but little more. No one forecasts junk or garbage, because it just doesn't seem very important. But it is.

8

Communist Junk: The Special Case

Junk is closely related to political systems, since the typical junkman is a free enterpriser of the wilder sort. To be a good junkman, you have to understand supply, demand, and free markets in a way that few businessmen do, and you may have noted that we have talked a lot about enterprising spirits in the business. Indeed, if you favor no government controls, free enterprise, minimal taxation, and all the rest of the capitalist system in its purest form, junk is for you.

But there are lots of communist countries where such free enterprise is discouraged. This raises a problem: in a communist state where no one cares about junk, who handles it? Moreover, it is common for communist countries to consider key data as state secrets, among them junk information. It is hard enough to find out anything about junk in an open economy, but impossible in a police state. After all, who will write glowing odes to piles of trash or, for that matter, even admit that such things exist? It is all probably some capitalist/imperialist plot since, in communist utopias, there is no junk.

Hence this chapter lacks the precise research of the others. It just isn't possible to explore piles of rubble in any communist country as yet. Hopefully, with the gradual relaxation of the Cold War, we can look forward to long and delightful hours reading communist junk in the near future, but for the moment this chapter is more hypothesis than proven fact.

Poor Communists, Poor Junk

A few key points seem very reasonable. Red China, North Vietnam, and North Korea are very poor countries, and it is very likely that they are still in the everything-gets-used category. These countries have large amounts of available unskilled manpower, and they need everything. So one would expect to find the honey-bucket carriers, the trash-pickers, the use of old tin cans, and all the rest. Indeed, given Red China's ability to organize huge numbers of laborers for almost any purpose, one expects that such trash-picking would be quite efficient. Visitors to Red China have often commented about the very clean streets in cities, and the neatness of the communes. Such cleanliness implies that either there is no junk generated, or that whatever is generated gets picked up right away. The absence of flies and bugs also suggests that whatever is being done, it is done cleanly. The usual dirt and dung associated with poor countries seems to be absent.

More affluent communist states present other problems, however. These countries' income is already above the piling-up phase noted in the last chapter, yet they do not have the private junkmen such countries have. And they also have somewhat different economic patterns, which would lead us to expect a somewhat different junk pattern. So, if we took a close look at the Soviet Union, plus Eastern Europe, what might we find?

Scarce Goods, Scarce Junk

The first major difference between these communist states and the West is that consumer-goods production is so very much lower for the given income than in the West. Cars are very scarce, while other durable consumer goods such as TV sets, refrigerators, stoves, and similar items are also hard to come by and quite expensive. Lots of trivia, such as electric carving knives, home driers, and hairdryers may not be produced or available. And even soft goods, such as clothing, are not typically all that plentiful, particularly as style goods. That is, you don't find fad items, such as bell-bottom pants, in common consumption.

The absence of cars means two things for junk. First, there aren't many old cars around to be junked, and hence auto wrecking yards must be rather scarce. Second, people can't be very mobile, in the junk sense. Remember, as we discussed in Chapter 3, if you want to live on junk in the United States, the first thing you need is a car or pickup truck. It would be hard to bring home a discarded sofa or freezer on the bus (it can be done—I've seen it happen in the piling-up type of countries, but not too often). One occupation that has to be scarce in the communist countries is that of private scrounger.

The Junkyard: A Capitalist Institution

One also wonders how state-owned junkyards would operate. This is exactly the sort of light industry that planners have very great trouble with in all communist countries —remember how complex the supply/demand patterns for auto wrecking yards are. Any respectable planner trying to figure out the total supply/demand situation for all umpteen million used auto parts for the next five years would either be wrong or insane within the first three days. You just can't plan such things very well.

Hence the forecast is trouble. One reads that in the Soviet Union, you take your windshield wiper blades with you when you leave your car, since they get stolen all the time. One also hears from time to time from the Soviet press about great difficulties in parts supply. It apparently is very difficult to find the right spare parts for the relatively few Soviet cars (there are only about 2 million cars in the whole country, compared to over 85 million in the United States). Could it be that all that auto trash is actually performing some critical economic function? After all, there are not enough mechanics in the United States to keep 85 million cars and 20 million trucks and buses running, yet they do run. Someone has to be doing various unrecorded things out there. But in a totally planned economy, where an entrepreneur cannot start a junkyard, such things cannot happen, and the mainstream gets into trouble.

The scarcity of consumer durables must also mean that very few of these get junked. Repairmen are a real problem in the Soviet Union and Eastern Europe, since this also is a light industry done by small privately owned shops everywhere, and organization of government-owned systems tends to be difficult and expensive. Most communist countries do allow for some private enterprise here—individual citizens can do repair work, often in their spare time. But parts supplies are a problem, since the state-owned firms that make durables are typically rewarded for output of total units, not parts. As a result, if any parts are available, they get assembled into total units, even if the clever planners include them.

Blat and Five Percenters

At this point, the *blat* man enters. The *blat* man is what we would call a five percenter—for a price, he can fix things up. He is a visible and widely commented-upon figure for communist industrial production and firms, but he must also be very active when you need some little, hard-to-find gismo for your TV set. For a price, he can get it.

In junk, since supply/demand relationships determine what happens in the West, prices tend to be very realistic. If a 1965 Chevy transmission is in demand, price will rise, junk dealers will note the fact, and they will make sure that whenever they run across one, it gets taken out of the car and put where someone is willing to pay for it. But in the Soviet Union and Eastern Europe, prices for all industrial goods are set in a variety of planned ways that virtually guarantee that they will not reflect true values. Hence there is no feedback system to tell anyone what to save and what to throw away as scrap. And since there are very few profit-motivated junkmen around either, one expects that all sorts of problems arise. Even major Western firms use old stuff from time to time, when it pays to do so—in the Soviet Union, it is hard to imagine how anyone would, at least in any rational way.

One can see the difficulties in this junk situation by reading utopian literature about the brave new world. One can find endless discussions about economic planning for new industries and goods; one can figure out how society is to be reconstructed, with some planner or the other doing all the noble work. But nowhere in this vast literature will one find any discussion of junk or garbage. Such things presumably don't exist in utopia at all. Hence when the revolution comes, no one is bothering to plan such trivia. This is all right for a Red China, where everything gets used, but when the country becomes as wealthy as the Soviet Union, real problems arise. No one has even thought about what to do.

This problem is compounded by this free-enterprise nature of junk. The stuff just isn't suitable to homogeneous planning and careful long-term organization. Hence we can expect that wherever junk serves some useful purpose in society, as in supplying spare parts for repairs, the problem will be large and difficult in the communist society. Scattered references to the difficulty of repairing cars and consumers durables, troubles with repairmen and their organizations, and similar complaints found in the Soviet press, do suggest that the junk problem is being handled rather badly, if at all, in that country. Eastern European countries are somewhat more willing to let small private repairmen do the job, but even here, fixing things is very difficult.

Junkies' Export Contributions

The Soviets also have great difficulty selling industrial and consumer goods in the free world, even at very low prices. A major reason for their problem is spare parts—somehow they have never quite figured out just how to keep things going, and foreign customers, after their first disappointment, don't reorder. Junk has a lot to do with the success of the free-world countries, since where junkmen operate widely and know what they are doing, somehow the manufacturers also know how to provide spares.

You will see many examples of cannibalization of equipment where Soviet material is involved. This is the using of a brand-new item for spares. Sitting over in the corner will be a partially dismantled truck or tractor, which is used to provide parts to keep the others running. In difficult service conditions, I have seen as many as five cannibalized vehicles for seven running ones. This is a very expensive way to solve your parts problem, since some things are really in demand, like electrical ignition parts, while other things, like truck cabs, are almost never used. You pay for them anyhow.

Americans and Europeans rarely cannibalize, although it is common in poorer countries where junk items are rare and parts knowledge also is imperfect. You can see it occasionally in military situations, where the planning process resembles the Soviet Union—no one is sure just what parts are needed and when. It was extremely common during World War II. But when you see cannibalization, you are looking at real industrial inefficiency, which is where the Soviets tend to be. It takes 8 or 9 million dollars of investment in the Soviet Union to get an extra million dollars of output annually—Western Europeans or Americans can get the million with perhaps 4 million dollars in investment. And smart junkmen make some modest contribution to this efficiency.

Industrial Junk

Industrial junk is a mystery in the communist countries, mainly because such things are state secrets. Very few foreigners ever get to a Soviet or East European factory, and even fewer are good junk readers. A couple of possibilities exist, however. One is that lots of junk and scrap really isn't. It gets sold through *blat* men to other factories who just happen to need a few coils of copper wire, or some sheets of steel. Since few planners really know much about what is needed for any industrial process, theft possibilities are extensive. Remember they are pretty good in the United States too—from time to time, some employee is arrested for carrying out some very good junk or scrap, such as circuit boards, transistors, or brass tubing. One can only wonder what goes on in a world of shortages, inevitably defective planning, and unrealistic prices.

A second possibility is that, given very unrealistic prices for many items, some kinds of industrial scrap gets produced in abundance, while other scrap is virtually nonexistent. Suppose that for any reason, aluminum sheets are very underpriced, in terms of what they really are worth, while labor is fairly costly, and brass is very expensive and overpriced. One would expect that there would be lots of aluminum scrap, if factory managers were given quotas to keep costs down (as they sometimes are). The system would work so that processes that saved labor and wasted aluminum would be used.

On the other hand, there would be almost no brass scrap, since its presumed price is very high. And note how this percolates down the system. Because aluminum is very cheap, the scrap piles up, and it may not even be collected. But suppose it really *isn't* cheap? Then there is a shortage of aluminum, and some planner has to figure out how to ration it.

This business of relative prices and costs is exactly what goes on in the West, except that the prices are real—they do reflect true scarcities. If the price of copper rises very fast on the world market, one can note, with very little time lag, great interest by junkmen in getting scrap supplies. Huge piles of automotive wiring harnesses that used to be around auto junkyards disappear—the owners, seeing the higher prices, find it worth their while to burn off the insulation and ship the wire to recycling mills, where it is melted down for reuse. If copper prices fall rapidly, the scrap, particularly the somewhat hard to get stocks, will pile up again. But in the communist countries, the price mechanism doesn't work this way, since prices are set artificially, based on vague criteria (often historical—some prices were set in 1933 or 1941 and haven't been changed since).

One can feel sorry for the communists on this point, since it would be indeed surprising if they managed to get their industrial scrap system working at even 50 percent of the Western efficiency level.

But the Soviets and Eastern European communists are very uptight, string-saving types. Operating in scarcity economies, they constantly worry in their press about saving everything. Labor is carefully controlled, and costs are somewhat unimportant, so one would expect to find considerable salvage efforts in state-owned firms and in government-controlled garbage systems. And indeed, we can observe such activities in and around communist cities. The streets are very clean (paper and other garbage is scarce, for the same reason it is scarce in poorer countries—not too much gets produced to feed into the disposal system). The Soviets in particular have had an unbalanced labor force ever since World War II, when so very many of their men were killed, and one way to use unskilled women is as sweepers and streetcleaners. Many visitors to the Soviet Union for many years have commented on such work being done by women.

One suspects that the garbage is of low quality, given all the above. Rarely would there be a discarded TV set or consumer item in the garbage, and any kind of scarce consumer good would be repaired and reused endlessly. Given fairly continuous food-supply problems, it is doubtful that the garbage would include many edibles either. But it probably gets picked up very efficiently.

In spite of the junk difficulties, Soviet citizens somehow stumble on. One reads of bootleg rock records made on discarded

X-ray film—now where did the bootleggers manage to get that film? Occasionally someone discovers an underground document or book mimeographed and passed around—the problems of getting hold of a mimeograph machine in a country like the Soviet Union must be enormous, but somehow it gets done. And finding enough paper to run off the copies is also a problem.

Somehow most of the privately owned cars seem to run too. In spite of parts problems, citizens manage. One expects to discover a vast underground free-enterprise system, which will do things for a price, and apparently such a system operates. A similar system has been widely discussed in the Soviet press for industrial materials, so there is no reason to expect that it wouldn't happen in consumer junk.

Free Enterprise and Junk

Those who are fond of free-enterprise economies might enjoy this odd dimension of the communist countries. Planning an entire economy looks easy, until one has to do it. Then the question of what to do with the literally millions, or even billions, of bits and pieces, both new and used, needed to run a complex industrial economy catches up to the planners. They just can't do it. To plan properly for trucks and cars, one must forecast precisely how many of each part (over 15,000 per vehicle) will wear out, break, or fail in the next five years. Even the most powerful computers ever built, or even likely to be built, are not good enough to do this job well. Let's face it, they can't do it at all.

Meanwhile, back in the West, we observe that the lowly junkman and scrap dealer help out the whole industrial system by supplying, for a price, all those bits and pieces that people seem to want. They do it so easily that no one ever pays much attention. And because they are so good, the firms that make the new stuff have to be good too. How can you sell a customer a new windshield wiper motor at a fat profit if he can go over to Joe's Wrecking Service and get a good one for half the price? The way you do this is to have a superbly organized parts system and after-sales service organization. So, firms do this because they have to.

Soviet firms, lacking this competitive pressure, plus any knowledge of what is really involved in junk, find that they can't compete in free markets. Their government worries about it, but somehow the essence of the problem eludes their grasp, and forty-five years after the revolution, they still can't do the job right. It's nice to know that the junkman, auto wrecker, and trash scrounger are doing their part to keep the world safe for capitalism, even in the communist countries.

Soviet and East European firms do have massive pollution problems. This stems from their close copying of Western industrial techniques with some time lag. If American steel mills used to belch huge black clouds of sulfurous smoke, then the Soviet firms will, too. If an American chemical company dumps crud in rivers, you can bet that the Soviet firm will be doing the same thing, with the same dismal results.

In the West, however, citizens can scream about such things, and they do. It is much more difficult in the Soviet Union, in part because the typical citizen does not see the plants. Remember, few have cars, and plants are not easily accessible. You may see a black cloud on the horizon, but you may never get close enough to really figure out what it is. And without cars or much light industry to provide things like fishing gear, few citizens get to remote rivers, only to find out that all the fish are dead. So the problem tends to stay in the background. Having a controlled press always talking about your country's virtues doesn't help much either.

Hence the few perceptive and knowledgeable observers who do know what is going on, or suspect it, have noted that communist industrial pollution is at least as bad, if not worse, than the Western style. And the classical citizen/political feedback mechanism to force someone to do something about it is lacking, so reform here is likely to be slow in coming.

We are back again to the worker as hero, if he is making some new industrial product. Who ever saw a junkman in a communist political poster? If all your heroes are producers, and no one worries about what comes out, then one can expect problems. And junk problems, the communists apparently have. The thing that saves them, so far, is the relatively low level of available consumer goods, including such things as wrapping paper. If little goes in, very little comes out, ever.

9

The Age of Affluence

This book got started in 1970, when I found myself with an hour to spare in Fuji, Japan. I wandered along main street, off into the smaller alleys, away from the bustle of that pleasant provincial city, and there, surrounded by high cement walls, was a junkyard. The gate was open, and several workers were busily stacking some materials into a truck.

But it was all wrong. I glanced at the various sorted piles of junk. There was a pile of 1-inch copper tubing ends, about 16 inches long. No crafty Oriental would ever have planned production so badly that he ended up with a thousand pounds of pure copper pipe in short pieces like that, but there it was. Over in the other corner were a half-dozen machine tools, lathes, rusty and forlorn. Machine tools are the ultimate in industrialization and economic progress, yet here were a group of perfectly good machines, obviously used, but clearly not too battered or worn. Remembering my dad's instructions about lathes, I took a close look, and sure enough, all the gear-cutting gadgetry was manual—these were old models, suitable for cheap machinists and lots of hand labor, but not too good for high-volume, low-cost production runs.

The men were loading scrap iron, and I could make out perfectly good engine blocks, chunks of grimy sewer pipe that weren't even cracked, and an occasional toilet bowl. It was prime scrap, neatly sorted, and it would bring a good price from any cast-iron user, but the stuff in the load was just too good to throw away. I remember tales of my friends who had seen Japanese kids eating out of garbage cans right after the war, and it was hard to believe that only twenty-five years later, Japan could be so opulent that perfectly good chunks of sewer pipe could be casually tossed back in the melting pot to be recycled.

And right down the street from the junkyard, a half a block away, was a 2-foot bundle of cut-off copper wires from a rewiring job on the telephone pole above. Lying right in the gutter, with no one picking it up! Copper was 40 cents a pound, and there must have been four or five pounds of perfectly good wire there. I watched the wire and the junkyard, realizing that somehow everything I thought I knew about Japan was all askew, yet not quite understanding what it was. And that wire was still there the next day, when we drove down the street to leave town. I almost picked it up myself, but I couldn't quite figure out what to say when I came back through American customs in San Francisco.

I suppose that nothing is ever quite what it seems in Japan, but the sight of so much high-quality junk in the country made my mouth water. It also led to some thinking about what was going on, and this book is the result of that particular chain of thought.

Affluent Country Socio-Economic Characteristics

As countries become more affluent, their junk problems shift dramatically. Here we are considering what is now the richer part of the world, with the exception of the United States and Canada. This group of richer countries includes most of Western Europe, Japan, Australia, New Zealand, and possibly the white sector of South Africa.

Such countries have per-capita annual income of from $1,200 to $3,000 per year. Since World War II, these countries have led in the growth race, increasing income faster per capita than almost anyone else. Gains of 5 percent per year are common, and Japan has averaged around 10 percent. At this rate, real income would double in a bit over six years. Americans riding a 2- to 3-percent income-growth curve can only ponder what it would mean to them, if American growth had been as fast. A typical 1946 schoolteacher earning $2,000 per year would now be making over $35,000 per year.

Some countries, like Great Britain, have not grown as fast as the rest (around 2 percent per year), but for most countries in this bracket, growth is expected and accepted. From a group of war-torn and war-weary economies in 1945, barely able to feed, clothe, and house their populations, these countries have come to the point where they can seriously expect to catch up to the United States by 1985 or 1995. Whatever else they may be doing, they are moving very rapidly into the affluent society.

The way you get rich is to produce things efficiently, so these countries have a full range of power plants, steel mills, chemical complexes, plastics plants, paper mills, electronics industries, auto manufaturing, and all the rest. Not every country has all of these, but every one does have quite a complex industrial establishment. Only New Zealand, which has managed to get rich with specialized agriculture, does not have the really extensive industrial development characteristic of this group.

These countries also have lots of dirty coal mines, iron-ore operations, oil fields, and similar extractive industries where the raw materials are found. Often this development is very old, since many of these countries were in at the very beginning of the Industrial Revolution. Indeed, England started it all. If it hadn't been for two world wars in this century, which set back all of Europe and Japan for many years, this chapter would be quite different, since it is quite possible that the countries would not be as rich or richer than the United States.

High productivity means high pay, so these countries have workers who are paid better than any in every place but Canada and the United States. In a few really productive industries, such as Japanese shipbuilding, pay may even exceed American levels, but generally it runs from half to two-thirds as much. The upper-middle-class professionals, managers, and administrators often make as much as Americans, although average income is somewhat less.

Population Growth: The Un-Problem

And the success of their economic development has led, in many countries, to labor shortages. This is the part of the world where population problems are very small—it will take something like 150 years to double the British population at present growth rates, and in none of these countries is the population growth rate over 1 percent per year. Birth rates were higher eighteen to twenty years ago, but not by much, so each year a smaller number of young people come into the labor force. And there is a big hole in the population of many countries (Japan, Germany, Britain) in the forty-five-to sixty-five-year-old bracket. This is the generation that actually fought World War II. The result is labor shortage in all but a few countries. Some, like West Germany, solve the problem by importing Mediteranean types to do the dirty work, leaving the "good" jobs for local citizens. As we shall see, this has much to do with junk, since trash collection is always a low-status job.

These countries, in one generation, have come from rather poor countries, able only to provide minimal standards of food, clothing, and shelter for most people to very affluent countries. And, like all others who have come the same way, junk and its problems were never even thought about. Hence, as rapid economic change occurred, so junk problems grew. And even now, no one really knows what to do about it. It all happened too suddenly to adjust to.

Pollution problems have been around these places for a long time. Remember, this was the area where the Industrial

Revolution began, and before 1850, Great Britain was having trouble with air and water pollution, to say nothing of the problems of sewage disposal. The dark satanic mills, with their pinched industrial proletariat and smoke-belching chimneys were a British invention, to be copied eagerly by all countries in this group and the United States. And since, until very recently, no one cared much about pollution, problems grew steadily worse. Americans worry about cleaning up their landscape, but much of industrial Europe and Japan is much worse. Entire rivers die in Europe, as poisonous substances are dumped into them by factories; smog was known in the Ruhr before the word was invented, coming from soft-coal smoke used in the mills; and the oceans around major cities often are cesspools of rotting raw sewage and garbage. Such problems are beginning to be recognized, and much is being done to eliminate them; but much, much more remains to be done.

This part of the disposal problem is being helped dramatically by the discoveries of major supplies of natural gas in the North Sea. England is already well on its way to moving from the soft-coal standard to the natural-gas standard, and one of the first casualties of this shift were the famed London fogs. The air is a lot cleaner now that soft-coal particles are not everywhere, and there is every prospect that the cleanup will continue. Even fish are returning to some British and continental rivers as industrial plants are gradually cleaned up, and as sewage treatment facilities improve. But this phase of the effluent problem is getting all the attention at the moment. Junk in the way we have been looking at it has not yet got the recognition it deserves.

Beyond the Minimum—And Still More Junk

So everything is up in these richer countries, and the age of affluence is dawning. Minimal human needs have been met everywhere, and as incomes continue to climb, citizens eagerly begin to accumulate the now traditional consumer-goods package. They buy their TV sets, washing machines, refrigerators, rice cookers, and automobiles, and count the days until they can afford more. They also buy much better housing, because the old cold-water workers' flats are not good enough for the affluent younger generation. And, as they move toward still more affluence, they are creating a whole new series of junk problems, which will plague them for quite some time to come.

Junk Tour of Europe

So you decide to take that tour to Europe at long last and stay long enough to catch the local spirit. How does it all look?

One thing immediately obvious to any American is the traffic. Autos have come to stay, and there are many more of them per square mile (though not per person) than in the United States. Thirty years ago, one Englishman in forty owned a car, and everyone took the train. Now, about one person in five owns a car, and they all seem to want to be in the center of town at once. Monumental traffic jams are a fact of life in every European country. Japan is just as bad, in part because roads don't get built as fast as the cars do. Older Americans, remembering similar momentous traffic tie-ups in the 1930s, can nod wisely. It all happened here too, and not so long ago.

If you look, the roads are getting built. Some tricky city intersections are being restructured; one-way streets are coming in fast; parking high-rises are being stuck up near the center of town; and the freeways are stretching across the country. Of course everything is jammed as soon as it opens, because the planners are ten years behind the motorists. It is surprising how slowly any planning governmental group responds to the obvious. A few more cars come in, and thinkers note that this is just a passing thing. Everyone will go back to the buses and trains in a year or two, as they should. So the money goes into new passenger trains. But the motorists don't go away—they expand very rapidly. When the total crisis point is reached, someone finally begins to do something, but by then it's too little and too late. No one really knows what the saturation point may be, since even the United States hasn't got there yet, but it clearly is a lot higher than where Western Europe and Japan now are. There are 106 million motor vehicles of all sorts in the United States for 207 million people, and the number still rises, even as population growth slows down. Europe is not halfway to this point yet on a per-capita basis.

The car explosion is followed in about eight to ten years by the junk car explosion, and of course no one bothers to forecast that either. The cars wear out, and no one knows quite what to do with them. Remember that only ten or fifteen years ago, junk cars were few and far between, and the few junkyards stripped cars as clean as they do in the piling-up types of countries, for exactly the same reason.

But suddenly the whole junk business gets out of whack. A few years ago, if an Englishman wanted a used rear end for his 1948 Morris Minor, there were only a handful in the country, and the junkies knew where they were. Now not too many people want such things, but the wrecking yards are loaded with wrecks. Parts prices decline to nothing, the wreckers fill their small yards quickly (remember how scarce land is in most of Europe), and after a while they refuse to take any but the late-model wrecks. In short, they start to operate like the American junkyards discussed in Chapter 4.

Junk Tour of Europe

Labor costs are going up fast, and remember all those labor shortages—few people want to work in a wrecking yard when other, more desirable jobs are available for the asking. So it gets very expensive to take off all the little bits and pieces that might be sold. Besides more affluent customers are beginning to buy new parts from the manufacturers, and moreover, the wealthier citizens are not fixing, they are trading in—and up, in most cases. You start in Europe with a Morris mini, Volkswagen or Fiat 600, and with luck and more income progress up to a Jaguar or Mercedes. Since land is very scarce in Europe, car twinning possibilities also are scarce. Remember, this is where a lower-income person keeps one junker twin to his running car. It works in the wilds of Queens or Indiana, but not in the working-class parts of Birmingham or Rotterdam. There just isn't the room. So even this minor demand is missing.

Where are those more desirable jobs? Well, it seems that Ford of England pays very well for production-line labor, to bolt up the new English Fords, and they pay still better for engineers, toolmakers, maintenance technicians, and all the rest. And European cars may be even harder to take apart than most American ones. In short, it's easier to build new than to fix, so the useful life of cars drifts down to eight or nine years, and here we are, with piles of old junkers and no place to put them. And, in their more or less complete condition, they make lousy scrap material too. As in the United States, there is too much copper, plastics, fibers, and zinc mixed in with all that good steel.

Given labor shortages, ease of making new ones, space shortages, and rising wrecking costs, many European countries find themselves getting into the abandonment phase for cars faster even than the United States did when its auto population was the same size per capita as Europe's is today—which was around 1929, actually. So the police and sanitation departments, to their dismay, find themselves beginning the familiar task of hauling in the junk. What do they do with it? Sell it off to the junkers, if possible, for one thing. They also are beginning to try ingenious new ways of disposal, such as dumping them in the right spots in nearby bays to make fish shelters. Indeed, what to do with the old cars is rapidly becoming a preoccupation of many governments. So far, no really imaginative solutions are in sight, and of course the problem will get a lot worse before it gets any better. Remember all those cars to come.

Bring Your Own Shopping Bag

As you shop in Europe, you come across a few odd notes for an American. Many of the shops still don't wrap things—you are expected to bring your own shopping bag. Remember that a high percentage of all garbage is paper, and count your blessings. In these countries, in part, paper is still scarce enough and expensive enough not to get casually tossed into use just once and then discarded. Smaller shops may still use newspapers for such items as fresh meat or fish. The pattern is very mixed, since various countries are at slightly different stages, and in the posh shops you will probably find the American pattern. But out in the provinces, older patterns may prevail. If they do, wait five years and come back. By then, everything will be wrapped and packaged, and then put in a bag (or maybe two), American style. And about the same time, you will be reading about how some city government somehow has filled up its garbage dump and is desperately seeking space for more dumping. This is a big issue in Toronto, where I am writing this, and they are right on schedule. In five to ten years, the Europeans will be where the Canadians are now.

You will note that in most cases the bottles you buy are returnable, and you pay a deposit for each one. It is hard to remember that in 1955, this was the American pattern too—disposable containers, except for the classic tin can, are really very recent innovations. But this too will change. Now the kids eagerly seek out the old bottles to return for credit; by 1985 they could care less. It just won't be worth it anymore, and by then the country will be using nonreturnable containers anyhow, as labor prices go up and materials prices go down. It takes a lot of hand labor to handle returnable bottles, and when wages get over two dollars an hour, it's just too costly.

Import Your Friendly Dustman

In France and West Germany, the pattern may go on for much longer. The reason is simple—these countries (along with Switzerland and the Netherlands) are willing to import cheap, low-grade labor from poorer countries to do the dirty work. Such immigrants, who come alone, without their families, are the garbagemen, the assembly-line workers, the cleaners, and all the other dirty jobs. They get relatively low pay by German standards, but quite high pay relative to incomes in Turkey, Algeria, Portugal, southern Italy, or Yugoslavia, which are still in the piling-it-up phase of junk development. And because the immigrants are willing and anxious to come, the dirty work gets done very well. In England, a trickle of black immigrants from India, Pakistan, Uganda, Kenya, and the West Indies came in to do the same work, and more came in ten years ago, but the British are unwilling to bring in masses of black people to do the work. Hence, we can forecast that for the time being, West Germany and France will be reasonably clean, while Great Britain will get scruffier.

Americans can see this point by pondering how our junk would disappear, if we let in perhaps 10 or 15 million very poor Asians and let them work at less than minimum wages. They might do unsavory things like eat out of garbage cans, but they also would enthusiastically clean up all garbage and junk, including all those old cars. We won't let them in, so the junk stays, as it will in Japan and Great Britain, who generally have similar immigration policies. Remember that garbage and junk is always handled by an underclass, and one of the problems of affluent countries is that the underclass gradually disappears. When it does, you have a trash problem, and this is where the European countries that don't allow much lower-class immigration are going, sooner than they think.

Because this underclass is still around everywhere in Europe, you don't see much visible trash, and what there is tends to be relatively low-grade stuff by American standards. Appliances are still expensive and scarce, and labor is still considerably cheaper than in the United States. Hence you won't find, even in most upper-middle-class neighborhoods, the old TV sets, bed springs, and all the rest of the discarded consumer package that you will in America. What is discarded still gets picked up very quickly by the maid, or some junk peddler of the old school, to be fixed and resold in some dreary used-furniture shop in the working-class neighborhoods.

Edging Towards Affluence

These affluent countries are still retreating, ever so slowly, from the spacial patterns common in the piling-up countries. That is, the inner city is still full of fine apartments, interesting homes, and other signs of affluence in a world without cars. But as you wander around Europe, and even Japan, you can see the slow shift to the North American style. The suburban explosion has begun, just as it always does as people get cars, TV sets, home appliances, and telephones. Why live in a crowded city apartment, where the kids can't play, when you can just as easily live in a single-family house at the edge of town, or maybe even farther out? So move out of the center of London, or Paris, or Manchester, along the main rail lines, and what you see are suburbs that resemble those near major American cities. The houses look the same, have the same implicit life style, and are snapped up as fast as anyone can get them built, even though they cost more than similar American ones.

This pattern is moving more slowly in Europe and Japan than in North America because of land problems. It is much more difficult to obtain good residential land near major European cities than in the United States, and it typically is very expensive. Hence to date only the elite have moved out, and European cities, to American eyes, look very lively—it is really time to travel. Manchester looks something like Cincinnati did around 1950, because the land-use patterns are just that way. But wait twenty years—it will all end.

Meanwhile, this more land-intensive pattern results in junk differentials. The high density encourages easy pickups of junk around apartments and flats, and labor is still cheaper. European countries have fewer resources, so carefully salvaged steel, tin cans, aluminum, paper, and plastics may have better markets for recycling, and they are closer to the source. The result is that now you see it, now you don't. The trash and junk is on the sidewalk for a brief period before it gets whisked away by someone, because it is both easier and more profitable. Moreover, the junk you see will be lower-grade stuff than at home.

Out in a very few posh suburbs, there is a hint of the new junk world to come. From time to time, for brief periods in very small areas, you can see and maybe even get the kinds of junk that a typical upper-middle-class American suburb yields—including the broken TV sets and old sofas. But I would not advise any American to try to make a living off junk, as you might in the United States. You would find yourself involved in competition with the local underclass who picks the stuff up for a living, and who would be upset to find a foreigner around. Besides, there are far too few areas exploitable and mineable for junk. Better go to Los Angeles—the living is easier.

But wait ten years—all the signs of progress are there. Books, magazines, and newspapers are becoming surplus commodity as their consumption explodes among a more literate and discriminating population. Old clothes, which used to find easy and profitable use as workingmen's gear, or industrial rags and wastes, are now beginning to pile up, and the export of used clothing, once an American monopoly, is being broken as various European countries find they have enough to send to Asia and Africa.

This stage is one where virtually all the gains in consumption lead to good junk later on. Until a country gets to affluence, any gains in income tend to get used up in food, basic clothing, and better shelter. When these things are provided for the great majority of the population, the next step is to buy more stylish clothes, prepacked foods, consumer durables, autos, and all the rest. And as this stuff wears out or goes out of style, it makes excellent junk. Rising labor costs and difficulties in repairing things send stuff to the junk pile faster. Most of the countries discussed here haven't quite got to this stage yet, but they will. The new stuff is either in the system or getting in very fast, and all we have to do is wait a few years until it begins to wear out.

Of course no one is doing anything about this, or even seeing the problem. The reason is simple—most of these countries are still in round one. That is, the first new stuff ever built hasn't really worn out yet. Early TV sets in England are only five to ten years old. British sets have been around for twenty-five years, but there were so few of them in 1955 that the junkers really aren't very visible. Just wait until they begin to pour out the other junk end by the millions! The junk possibilities are mind-boggling. And better yet, every public agency will be caught completely by surprise when it begins to happen, just like the United States. In junk, no one ever learns but the hippies and junkmen, and they aren't talking.

Industrial Junk

Industrial junk in these affluent countries is exploding, both in quantity and quality. Where it can be dumped in a river or ocean it is; this is why pollution levels are so high. But where it cannot, the junkmen come into their own. This is the kind of world where a good junkie can make a fortune. Remember, the really good professional junkmen began back in the piling-up stage, which lasted a long time in Europe.

Beginning as early as 1900, such firms and individuals were active, and by now they are very well organized and managed indeed. Remember also that no one bothers to regulate, control, or even tax junkmen, so these firms and people have to be good—they are really in a competitive market, where only the fit survive.

But in the affluence stage, the growth of the business is phenomenal. Instead of a one-man firm with a beat-up truck patiently wandering around to five or six factories each day to get a few tons of scrap steel, the company now finds that it needs a fleet of trucks, leased railcars, and huge handling facilities to do the job right. Remember, all those new auto plants, steel mills, chemical companies, and so on are pouring out huge tonnages of good, usable scrap, all nicely segregated and ready to go back for recycling. Industrial scrap goes from hundreds of tons to thousands of tons per day of anything you would care to name. Each steel stamping means a few scraps of steel on the initial sheet—and now the company is producing a thousand cars a day instead of thirty. Each wooden packing crate made for an export order means a few scraps of wood—and now two hundred crates are made every day instead of seven. Each electric motor assembled results in a few bits of copper for salvage—and now production is 50,000 per week instead of 1400.

The result is that a few very competent junk firms get very rich in this phase, as they did in the United States during the similar development period. This is when junk dealers no one ever heard of come to the best restaurants in their Mercedes limousines with their bejeweled women. Such people are lots of fun, since they tend to have working-class origins, and often are of the wrong religion or race besides. Who else would mess with junk? Polite society is shocked, and intellectuals and social arbiters comment discreetly about the vulgarity of the new rich, but the fellow lighting his four-dollar cigar as he gently waves his chauffeur off could care less. Let the nice people figure out how to make money for themselves!

Rags to Scrap Iron to Poetry in Four Generations

The first generation collected junk in 1910 with a pushcart; the second generation did it in 1950 with an old truck; the third generation did it with a fleet of vehicles, an office downtown, and batteries of teletypes and secretaries. The fourth generation will go to the best universities at home or abroad, and probably end up writing poetry on Cyprus, or maybe managing a respectable firm that produces a new product. The daughters will manage to marry into very reputable, but unfortunately financially pressed fine old families. And, in 1985, the Japanese and Europeans will look around and discover, to their surprise, that they have a real junk problem. Relax—it's already happened in the United States, and by then we will have figured out what to do about it, if anything can be done. The Europeans and Japanese are welcome to this sort of technology and management.

Have you ever heard of anyone patenting anything about junk?

Junk Houses

This economic phase also is the one where the junk house problem begins to happen. All industrializing countries experience the same phenomena, which is the vast movement of farmers and peasants to the city. As this happens, the farms get abandoned. Note that this only occurs where birth rates are low—in poorer countries, many migrate, but, given high population-growth rates, many are also left behind to fight for scarce farmland.

In affluent countries, most people leave the farm to work in the new factories. Indeed, the factories may expand so fast that even after most people have left the farms, the country still has to import foreign workers. Back on the farm, the survivors tend to be very smart agri-businessmen, who know how to make the land very productive, and who are good customers for tractors, fertilizers, agricultural equipment, and similar items. In the United States, it is common now to find one farmer handling perhaps 2000 or more Kansas acres. In 1880 there was one farm family for every 160 acres. What happened to the farmhouses? Well, they got junked—or abandoned. The same thing is happening all over Europe, for the same reasons, at a somewhat slower pace. But the result is the same—junk houses. And these tend to hang around for decades, since there is not much of a market for them, and they gradually fall apart. Since they are out in the boondocks someplace, no one much cares except a few antique hounds who sack them for whatever may be left.

With luck, even here one can find some good junk. Americans dismantle barns to get the lovely old wood—Europeans sometimes take apart old farmhouses for the oak beams.

Back in the cities, junk houses also abound. As living standards rise, and people move slowly out to the suburbs, inner-city slums gradually get abandoned. So far in the United States, such housing has largely been filled up by Appalacian white hillbillies and black ex-sharecroppers moving into the cities to seek their fortunes. But we are experiencing a population slowdown, and we are also experiencing a continuation of a very old pattern, which is the move of the more successful ex-poor out into the suburbs. Even the blacks are doing it, and as they get richer, they will continue to do it. In Europe, unless immigration is allowed, there is no one left to fill up the cruddy housing, so it gets abandoned.

Once in a great while, some very wise planner sees that the best thing that could happen to the inner city is this sort of housing abandonment. So the property is bulldozed down, creating some very low-grade junk, and later a park is created. But far more often, such valuable properties are used for some new project. The land is too valuable.

New projects in cities cost very big money, and often land

clearance is started, but nothing is finished. That is, the junk just sits, often for years. Among the crud are all sorts of good junk, such as old oak beams, two by fours, piles of slightly tattered lumber, old bricks, lead and copper plumbing pipes, and much more. Such stuff tends to fade away from the site rather quickly, leaving nothing but the low-grade rubble junk, but often it takes quite a while to dispose of the good stuff. One can tell with considerable accuracy exactly how well developed the country is by observing just how long it takes to clean out the high-quality junk in such a project. If it disappears in two days, the country is still rather poor; if it takes six weeks to get rid of the good stuff, the nation is close to the total affluence level.

Because countries move very rapidly through this phase, take a good look often. It may have changed before you really realize what is going on. As one example, Japan, an admittedly fast-growing country, was at the everything-gets-used stage in 1946; by 1955 it had reached the piling-up stage; and by 1970 it was in the age of affluence. The way the country is going, it may well reach total affluence by around 1985. Almost daily junk changes occur, so one has to be reading junk all the time to make sure that nothing is missed.

What Went In Will Come Out

Most of Western Europe is well along toward total affluence, since lots of the basic inputs already are in the system—they just haven't started to come out yet. The cars, the consumer goods, and the packing materials are all in there someplace. Within five years, the most affluent parts of Western Europe will discover that they have a major junk problem. So far, most of the problem has revolved around cars, since these are big and visible, even when there really aren't too many of them around. Junk problems tend to accelerate when a country reaches the affluence level that West Europe now has, as the Americans learned to their dismay. But it could be worse—at least a large supply of very high-quality junk suggests that the country is well along in the wealth and culture race. Without good junk, your country is classified as developing, less developed, or poor. With quality junk, you're in the big leagues.

10

What To Do About It

Studying junk may seem improbable, but it's fun. While studies of modern junk are unusual, the idea really is not strange—after all, anthropologists have nosed around ancient junk piles for a long time, reading them to see what the culture really was all about. This study was nothing more than a modern extension of a very old idea.

You can go to almost any good museum and see lots of ancient junk, such as old shards of pottery, crumbling stone sculpture, rotting wooden beams from ancient houses, and cloth fragments. Most of this junk, while very old, really is rather low-grade stuff by modern standards, since all ancient cultures were at the everything-gets-used stage for millenia. But if a broken Babylonian pot suggests what kind of life style these ancients had, it is a reasonable and logical extension to take a look at a modern Plymouth door handle or aluminum beer can and similarly deduce how modern man lives. And, just as ancient junk differs somewhat from culture to culture, modern countries also provide quite different junk samples around the globe.

Junk and Nice People

A major reason why looking at modern junk seems odd is that it normally is not done by "nice" people. We have suggested that junkmen have very low status in all modern societies, standing only slightly above garbagemen and handlers of human excrement. Because the junkman's social position is so low, no one really cares much about the problem. It is far more interesting to study the activities of nobler professions. As a result, junk offers excellent possibilities, in affluent countries, for economic gains, and even freedom. No one cares— —so you can do what you want and be free of taxes, regulations, social discipline, and all the rest. Some perceptive people have spotted the free-life-style potential of living off junk, but as yet they are considered eccentric. Maybe they are, but a really expert junk user can do things that the rest of us uptight middle-class types can only dream about.

Because we don't care about junk, we have major social problems. The stuff seems to swamp us all the time. Actually, being swamped in junk is a symbol of total economic success, as the chapter on "everything gets used" and "piling it up" suggested. One can easily estimate accurately the power, prestige, and wealth of any country by sifting through its junk piles.

Junk Reading for Fun and Profit

The implications of reading junk are fun. If you learn to read it, you can structure your own operations in the country easily and correctly, since junk tells anyone who cares to look what is really going on. Politicians and administrators everywhere may make efforts to conceal or distort the truth, particularly if the truth is a bit unpleasant. But no one pays any attention to junk—it doesn't lie, nor does it get altered by zealous bureaucrats. Such individuals and organizations are incapable, given their own training and interests, of ever doing anything with junk. It is beyond their range of competence and interests. And for that reason, it pays anyone who really wants to know what is happening in a country to wander around seeing what the junk piles look like.

Hence one can discover more about mechanical skills, economic levels, and living standards by wandering through the nearest auto junkyard than by studying economic data for years. Just examine what is taken off the wrecks; study how thoroughly the hulks are stripped; indeed, see how many of them there are. The country's recent social and economic history is right there, ready for anyone to read.

In the United States and other affluent countries, junk depresses most people, and since the junk is a part of the pollution problem generally, it is possible that there will be considerable interest in getting rid of the stuff. Already some states have conducted vigorous auto-pickup campaigns, and many cities have tough laws about leaving wrecked cars lying around. Cities also have sponsored cleanups of junk-laden streams. As usual in such campaigns, the strategies followed are partial and ineffective, since they attack the junk problem in the wrong way. They start picking things up, when this really is the end of the line. If you seriously wanted to get rid of junk, the place to begin is before the stuff becomes junk, not after it already is littering the landscape.

Doing Something About Junk

So, what might be done, if we really got serious about junk? Here are some feasible, if politically complicated, suggestions:

We might begin by observing that junk accumulates rapidly when labor costs are high, when people are very affluent, when new goods are relatively cheap, and when price differentials between repairs and replacements are marginal. This is exactly where the United States is now. But we already tax and subsidize all sorts of things for noble social purposes; it would be easy enough to structure a set of taxes and subsidies to alter these relationships in favor of discouraging junk.

Many such ideas already are in use someplace, or at least proposed. Taxes on containers of various sorts have the effect of raising prices and possibly discouraging consumption. In Oregon, the law requires that all containers be returnable, rather than tossed aside. In Toronto, milk is sold in returnable plastic bottles. The results of such laws can be, if the price differentials are large enough, to discourage the dumping of many cans, bottles, and plastics.

Remember though that in a really affluent society, people may throw away the returnable containers anyhow. The cost has to be high enough to discourage this, along with encouraging the kids to scrounge the containers out of the garbage and return them for cash. But it can be done.

Cars can be made disposable in a variety of ways. One easy way is to tax new owners for costs of getting rid of old cars. If it really costs money to pick the junkers up, then the fellow who buys a new one might be made responsible for getting rid of the old one. Another thing not yet done, but which really has tremendous potential, is to design cars so they come apart with minimal labor input. If a junk dealer could *easily* salvage all those valuable raw materials, he would happily go pick them up for you.

One major reason why most countries outside the United States do not have much consumer-durable junk around is that the costs of new items are very high relative to repair costs. If manufacturers were required to design new items so that they could be fixed easily and cheaply, lots of junk would be eliminated. And if taxes were placed on new items, but not on spare parts and components, it would pay more to fix than to discard and buy new. Engineers sometimes gold-plate products, making them more complicated and harder to fix than they really should be. Certainly a technological society that can get to the moon or build 747 jets should be able to design an electric can opener that can be fixed for less than the cost of a new one! But note the sociology involved here. An engineer for General Motors is a prestigious, responsible citizen—the repairman is a greasy mechanic who lacks status. Who cares about him? Until we begin to see some attitudinal changes, the problem will remain.

The Return to the Repair Society

There is already plenty of evidence to suggest that we are drifting into a repair-oriented system again, more or less by accident. Many durable-goods retailers have discovered that if things can be fixed, you get lots of customers. This success, with effective after-sales maintenance has prodded others into moving in the same direction. A surprising number of consumer-durable manufacturers are beginning to push maintenance, because it pays. The maintenance manual that comes with the Ford Pinto actually tells the owner how to do simple repairs, and the car is selling very well. (It's also one of the easier cars to get into and fix—but not yet good enough. If Ford cares, I'll happily spend a few hours telling them how to do it better). Volkswagen also has pushed durability and easy maintenance, as has Volvo.

Another factor favoring fixing rather than buying new may be the anti-pollution laws. New cars are going to get much more expensive if the pollution-control gadets are actually installed, maybe as much as a thousand dollars more by 1976. If it costs you $5,000 for a new one, and the repairs are only $300, many customers, including some affluent ones, may be willing to fix. So far, the average age of cars in use has not risen, which would suggest that owners are hanging on, but it could happen. And if gasoline really gets to a dollar a gallon shortly, as is widely predicted, not only will smaller cars sell better, but older ones without pollution equipment may be kept to keep costs down. A 1931 Model A gets around 20 to 25 miles to the gallon, and 1965 small cars may get up to 35—they may be kept in use as a result. It really doesn't matter by whom, as long as the cars don't get abandoned someplace. The original owner may say to hell with it and buy another, but the next buyer, maybe a junkman, finds it profitable to fix up and resell.

Repair Hangups—Ten Million Unemployed?

One major reason why Americans get nervous about such suggestions is that if we really got serious about using things for longer periods of time, and if we really fixed instead of throwing away, we could have some major industrial employment problems. Most major consumer-durable-goods markets are close to the saturation level, and factories making new stuff are merely replacing the old. If that old junk keeps right on running, then where is your market? Of the 8 to 10 million cars sold each year in the United States, about 80 percent are for replacement, while the rest add to stock. But incremental additions to the American stock of cars is likely to decline to nothing within a few years, while if we kept the old ones running, perhaps only 2 or 3 million new cars would be needed each year to take care of the replacement of really old junkers. What happens to the auto-manufacturing business then? You could work through similar calculations for refrigerators, stoves, driers, washers, and lots of other consumer goods, and markets for new products could be in deep trouble if we really fixed things up.

When you count up the millions of assembly-line workers, foremen, designers, salesmen, technicians, managers, and others who make their living producing and selling new things, you may see why forced obsolescence and throwing things away is so important. If we really got serious about preventing junk, we would have to restructure our whole economy.

Back to World War II

But we did it once, when the United States was a lot poorer. During World War II, we put some 13 million of the very best and most productive young men into the armed forces, stopped making all the consumers durables and cars, and still got living standards up by almost 50 percent for the rest of the population during the war. The car salesmen, brokers, and all the other "unproductive" types went off to war or the shipyards, and people did fix old radios, cars, and what not. Incidentally, the quality of junk declined dramatically during the war—I was young then, and you couldn't find a thing in anyone's junk pile. People even fixed old dollar alarm clocks and eight-dollar radios, because they had to. Even the old tin cans got carefully flattened and dumped in special containers to be recycled.

But because getting rid of junk will require restructuring the economy, we are unlikely to do it unless we have to. And, though the present energy problems suggest that some costs will rise a bit, and some thing will be recycled, it does not appear too likely that this dramatic change will happen very soon.

Another way to get things restructured a bit, is to give some sort of subsidy to repairmen, while adding on costs of hiring men who make new things. We might, as one example, exempt repairmen who work in small business from their social-security tax payments—this would mean that such labor would be about 10 percent cheaper than other kinds. It would be that much cheaper to repair things in the future. But even though such a plan would lead more people into repair work relative to new work, it is unlikely to be done.

Moving Junk to Get Results

We could also shift junk from where it is, which is where

no one wants it, to where it is highly prized. Within the country, this is what we do, the hard way, when we begin with an abandoned car on a city street and eventually get it to a wrecker, then in its broken up state to mills where the scrap is recycled. We could also do it internationally—actually, trade in such things as scrap iron, copper, and old clothes is already very large and growing. But we could give the hulks to India or the Philippines, or some one else who knew what to do with them. As we suggested earlier, this would probably involve war, given modern attitudes toward junk, such a policy could not only make war more fun and positive, but also add both to income and education in many places that need it. In the end, such countries would advance along the junk cycle to where they could begin to pass along their junk to some other still poorer place.

Attitudinal factors make such thoughts impossible to execute in the real world. We think nothing of dumping lethal junk, like bombs, on helpless people, but nice people just don't dump nonlethal and constructive junk, like old cars, on poor countries. What would you rather have in the next war—junk in the form of bits of human flesh and bomb fragments, or junk in the form of an old Chevy, sitting invitingly on the beach?

The Junkie As Hero

Perhaps the most useful thing we could do is to get the junkman's social status up, so that younger people could get into the business without stigma, but neither I nor anyone else has any idea how to do this. Maybe we could have some TV documentaries, or Soviet-style posters of muscular, handsome junkmen performing nobly the world's work. Who knows? But such a social change could pay off big. Come to think of it, it might not. If junkmen were respectable, the government would begin to help them as they do other businessmen, and that means cartelization, control, more taxes, more regulation, and all the rest. No, maybe it's better to let the boys just quietly move along in their time-honored ways. At least the way it goes now, the junkmen can do their job without much fuss.

For other countries, it would be useful to see the junk problem before it hits. Anyone who has read this far will be able to spot exactly where his country is on the junk scale, and assuming normal economic growth, it is possible to forecast with considerable accuracy just when the junk problem will get more difficult. If you know it's going to happen, maybe you can begin to plan for it well in advance. And then, instead of being a crisis and impossible, the problem becomes only difficult. So far, no country has ever managed to do this, but stranger things have happened before.

Countercultures

The evolution of suburban living styles in the United States has generated a counterculture, which has a much healthier attitude toward junk. Indeed, the junk is necessary to the counterculture, since so much of it is based on the stuff. One lives in a junk house or apartment; one uses old stoves, beat-up sofas, mattresses scrounged from the city dump, and all the rest. Perhaps we are seeing some sort of symbiosis here, which leads to a self-correcting system. The more junk we generate, the richer we are. The richer we are, the more people reject the materialistic culture. The more people do this, the more junk is scrounged. In the end, the size of the counterculture is a function of the junk output of the mainstream.

Note that our sort of counterculture is only possible in an affluent country. There just isn't enough junk around to support a sizable counterculture elsewhere. You'd starve first.

But the more people who go counterculture, the better off we are in junk terms. If you don't like countercultures, try supporting the proposals made earlier in this chapter. The more efficient we become in commercial recycling and fixing up, the smaller the counterculture can be. If we really get efficient, as we did during World War II, then the whole counterculture disappears. A parasite can't exist without a host, and the junk piles of America, in all their wonderful and intriguing forms, *are* the host.

In the Long Run, Junk is Dead

In the end, all of this book will be irrelevant, since we do live on a finite spaceship earth, and before we are done, everything will have to be recycled. We happen to live in an odd period—odd because until just a few decades ago, everything got used everywhere. People were far too poor to throw away much of anything that had any possible value. Then a few countries got a growth track, which led to fantastic affluence, and now it is easy to throw things away. This may be the atomic age, the age of technology, or the age of affluence, but it certainly also is the golden age of the junkman. And, like so many other human things, it will end. Actually, it will just be different. Before many of us retire, we will be recycling things because we have to, and we will be giving junk and other effluents the respect they deserve. In a finite world, everything recycles in the end. So, if junk appeals to you, pay attention now, because it won't last forever.

So we end our meandering around the junk piles of the world. No one else cares, but anyone who understands what junk is all about, and how people use it, throw it away, recycle it, and generally adjust to it, is in a position to make money, see how the world really works, and have lots of fun. Try looking around—it can be interesting.

Also available from
Stein and Day

in hardcover editions:

A Great New Way to Make Money
  by Ralph Charell
The Joy of Money: A Contemporary Woman's Guide to Financial Freedom
  by Paula Nelson
How to Buy a Condominium by Lester and Patricia Brooks

in Scarborough Books paperback editions:

Starting Out: The Guide I Wish I'd Had When I Left Home by Lili Krakowski
A Woman's Guide to the Care and Feeding of an Automobile by Carmel Reingold
The Consumer's Guide to Banks
  by Gordon Weil
Extra Dollars: Easy Money-Making Ideas for Retired People by Ray Hoffman
Dollars on Your Doorstep: How to Run a Business from Your Home
  edited by Joseph Daffron
An Irreverent and Thoroughly Incomplete Social History of Almost Everything
  by Frank Muir

ASK YOUR BOOKSELLER!